MUSIC SEARCH

JUST ONE OF THE MANY AMAZING FEATURES FROM SYNC,
THE VOICE-ACTIVATED IN-CAR TECHNOLOGY AVAILABLE
EXCLUSIVELY ON FORD, LINCOLN AND MERCURY VEHICLES.*

SYNC. Say the word.

hands-free calling • music search • vehicle health report • turn-by-turn navigation • business search
911 Assist™ • real-time traffic • audible text • my favorites
Learn more about all SYNC features at syncmyride.com

*Driving while distracted can result in loss of vehicle control. Only use mobile phones and other devices, even with voice commands, when it is safe to do so.

Volume 20

Make:
technology on your time

For ages 9 to 99 — **Make: Kids**

ON THE COVER: Adam Savage, sporting his hacked Apollo space suit, photographed by Cody Pickens. Background illustrated by Adam Koford. Read an in-depth interview with Savage on page 32, and build the Hydrogen-Oxygen Bottle Rocket on page 90.

Columns

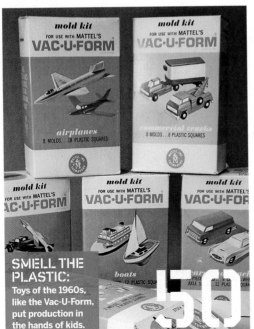

SMELL THE PLASTIC:
Toys of the 1960s, like the Vac-U-Form, put production in the hands of kids.

'50

Vol. 20, Nov. 2009. MAKE (ISSN 1556-2336) is published quarterly by O'Reilly Media, Inc. in the months of February, May, Aug., and Nov. O'Reilly Media is located at 1005 Gravenstein Hwy. North, Sebastopol, CA 95472, (707) 827-7000. SUBSCRIPTIONS: Send all subscription requests to MAKE, P.O. Box 17046, North Hollywood, CA 91615-9588 or subscribe online at makezine.com/offer or via phone at (866) 289-8847 (U.S. and Canada); all other countries call (818) 487-2037. Subscriptions are available for $34.95 for 1 year (4 quarterly issues) in the United States; in Canada: $39.95 USD; all other countries: $49.95 USD. Periodicals Postage Paid at Sebastopol, CA, and at additional mailing offices. POSTMASTER: Send address changes to MAKE, P.O. Box 17046, North Hollywood, CA 91615-9588. Canada Post Publications Mail Agreement Number 41129568. CANADA POSTMASTER: Send address changes to: O'Reilly Media, PO Box 456, Niagara Falls, ON L2E 6V2

Make: Projects

Hydrogen-Oxygen Bottle Rocket

Use electricity to split tap water into hydrogen and oxygen gases, then use this explosive gas mixture to power a two-stage, electronically timed rocket. By Tom Zimmerman

90

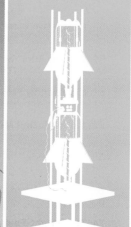

Auto-Phenakistoscope

Motorize a 19th-century parlor novelty, and keep its frames synched to an LED strobe by using a sensor and an Arduino microcontroller. By Dan Rasmussen

100

Lunchbox Laser Shows

Build three different laser effects machines that fit into metal lunchboxes to create exciting sound and light shows. By Mike Gould

110

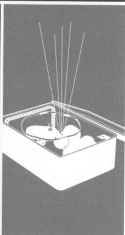

Are you up for a challenge?

What is the missing component?

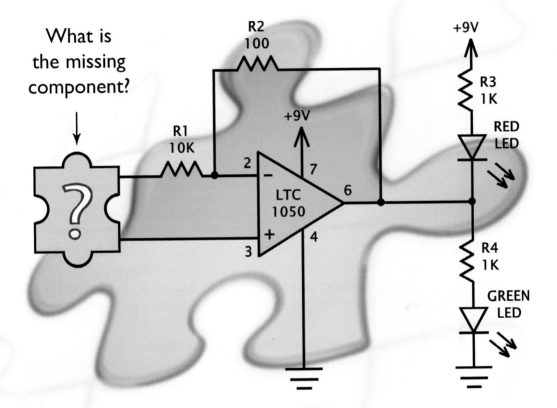

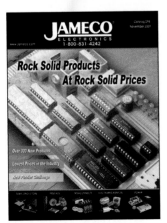
Industry guru Forrest M. Mims III has created a stumper. Video game designer Bob Wheels needed an inexpensive, counter-clockwise rotation detector for a radio-controlled car that could withstand the busy hands of a teenaged game player and endure lots of punishment. Can you figure out what's missing? Go to www.Jameco.com/unmask to see if you are correct and while you are there, sign-up for our free full color catalog.

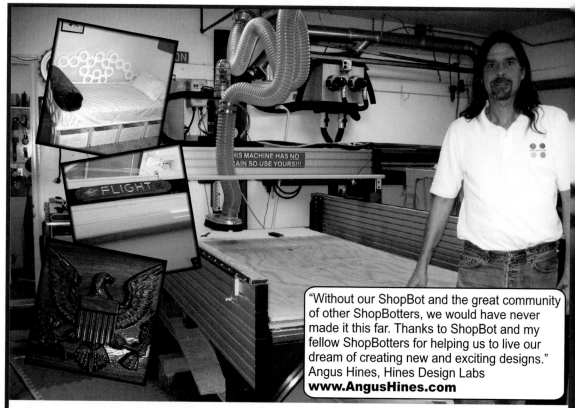

"Without our ShopBot and the great community of other ShopBotters, we would have never made it this far. Thanks to ShopBot and my fellow ShopBotters for helping us to live our dream of creating new and exciting designs."
Angus Hines, Hines Design Labs
www.AngusHines.com

Angus Hines' life as a Maker began at the age of two when he took his first toy apart to see what made it work. From that point on, if it had screws or could be taken apart and dissected, nothing was safe. He was bound and determined to see what made things work and find ways to make them work better.

After he retired from his career as a project manager and spent some time cruising the Chesapeake Bay, Angus began to get bored and restless. He turned to his first love of creating new things and purchased a laser engraver. Angus started Classic Marine Co. and began making instrument panels, indestructible LEDs and other marine Items, but, as business grew, he decided he needed a larger, more versatile platform. Angus purchased a ShopBot and expanded into larger design projects like furniture, small buildings and larger marine items. Not content to make things that already exist, he started another company, Hines Design Labs, and has now ventured into prototyping, design assistance and helping people take their ideas from a napkin sketch to a finished product. Angus' love of making has taken his company from a small marine manufacturing facility to a full-blown design prototype shop, and his ShopBot helped make it all possible.

What will you make today?

888-680-4466
www.shopbottools.com

Make:

Volume 20

technology on your time

Maker

GAME OF LIFE:
Michael McGinnis spent years making his Superplexus a reality.

DIY

1.2.1.

"Play is often talked about as if it were a relief from serious learning. But for children, play is serious learning. ... In fact, play is the real work of childhood." —Mister Rogers

Make:
technology on your time

EDITOR AND PUBLISHER
Dale Dougherty
dale@oreilly.com

EDITORIAL

EDITOR-IN-CHIEF
Mark Frauenfelder
markf@oreilly.com

MANAGING EDITOR
Shawn Connally
shawn@oreilly.com

ASSOCIATE MANAGING EDITOR
Goli Mohammadi

PROJECTS EDITOR
Paul Spinrad
pspinrad@makezine.com

SENIOR EDITORS
Phillip Torrone
pt@makezine.com
Gareth Branwyn
gareth@makezine.com

COPY CHIEF
Keith Hammond

STAFF EDITOR
Arwen O'Reilly Griffith

EDITORIAL ASSISTANT
Laura Cochrane

EDITOR AT LARGE
David Pescovitz

CREATIVE DIRECTOR
Daniel Carter
dcarter@oreilly.com

DESIGNER
Katie Wilson

PRODUCTION DESIGNER
Gerry Arrington

PHOTO EDITOR
Sam Murphy
smurphy@oreilly.com

ONLINE MANAGER
Tatia Wieland-Garcia

PUBLISHING

MAKER MEDIA DIVISION
CONSULTING PUBLISHER
Fran Reilly
fran@oreilly.com

ASSOCIATE PUBLISHER &
GM, MAKER RETAIL
Dan Woods
dan@oreilly.com

DIRECTOR, MAKER FAIRE
Sherry Huss

CIRCULATION DIRECTOR
Heather Harmon Cochran

MARKETING & EVENTS MANAGER
Rob Bullington

ACCOUNT MANAGER
Katie Dougherty Kunde

SALES & MARKETING
COORDINATOR
Sheena Stevens

ADVERTISING INQUIRIES
Katie Dougherty Kunde
707-827-7272
katie@oreilly.com

EVENT INQUIRIES
Sherry Huss
707-827-7074
sherry@oreilly.com

MAKE TECHNICAL ADVISORY BOARD
Kipp Bradford, Evil Mad Scientist Laboratories, Limor Fried, Joe Grand, Saul Griffith, William Gurstelle, Bunnie Huang, Tom Igoe, Mister Jalopy, Steve Lodefink, Erica Sadun, Marc de Vinck

CONTRIBUTING EDITORS
William Gurstelle, Mister Jalopy, Brian Jepson, Charles Platt

CONTRIBUTING ARTISTS
Roy Doty, Nick Dragotta, Julian Honoré, Jason Hornick, Adam Koford, Timmy Kucynda, Tim Lillis, Garry McLeod, Rob Nance, Cody Pickens, James Provost, Jen Siska

CONTRIBUTING WRITERS
Claudio Bernardini, Julian Darley, David Delony, Douglas Desrochers, Cory Doctorow, Nancy Dorsner, Sam Fraley, Mike Gould, Richard B. Graeber, Saul Griffith, Tom Heck, Ross Hershberger, Joel Johnson, Lisa Katayama, Patrick Keeling, Jeff Keyzer, Dylan Kirdahy, Bob Knetzger, Tim Lillis, Hayden Lutek, Thomas Martin, Michael McGinnis, Robyn Miller, Forrest M. Mims III, Dmitri Monk, Trudia Monk, Anton Ninno, Christie Noe, Jim Noe, John Noe, Terry Noe, John Edgar Park, Tom Parker, Joseph Pasquini, Michael H. Pryor, Dan Rasmussen, Gerald S. Reed, Adam Savage, Damien Scogin, Peter Smith, Bruce Stewart, Jeanne Storck, Peter Torrione, Gever Tulley, Matthias Wandel, Ben Wendt, Megan Mansell Williams, Tom Zimmerman, Lee D. Zlotoff

ONLINE CONTRIBUTORS
Gareth Branwyn, Chris Connors, Collin Cunningham, Adam Flaherty, Kip Kedersha, Matt Mets, John Edgar Park, Sean Michael Ragan, Becky Stern, Phillip Torrone, Marc de Vinck

INTERNS
Eric Chu (engr.), Peter Horvath (online), Steven Lemos (engr.), Kris Magri (engr.), Cindy Maram (online), Lindsey North (projects), Meara O'Reilly (projects), Ed Troxell (photo)

PUBLISHED BY
O'REILLY MEDIA, INC.
Tim O'Reilly, CEO
Laura Baldwin, COO

CUSTOMER SERVICE
cs@readerservices.
makezine.com

Manage your account online, including change of address:
makezine.com/account
866-289-8847 toll-free in U.S. and Canada
818-487-2037,
5 a.m.–5 p.m., PST

Visit us online:
makezine.com

Comments may be sent to:
editor@makezine.com

EVEN GREENER
MAKE is printed on recycled, process-chlorine-free, acid-free paper with 30% post-consumer waste, certified by the Forest Stewardship Council and the Sustainable Forest Initiative, with soy-based inks containing 22%–26% renewable raw materials.

PLEASE NOTE: Technology, the laws, and limitations imposed by manufacturers and content owners are constantly changing. Thus, some of the projects described may not work, may be inconsistent with current laws or user agreements, or may damage or adversely affect some equipment.

Your safety is your own responsibility, including proper use of equipment and safety gear, and determining whether you have adequate skill and experience. Power tools, electricity, and other resources used for these projects are dangerous, unless used properly and with adequate precautions, including safety gear. Some illustrative photos do not depict safety precautions or equipment, in order to show the project steps more clearly. These projects are not intended for use by children.

Use of the instructions and suggestions in MAKE is at your own risk. O'Reilly Media, Inc., disclaims all responsibility for any resulting damage, injury, or expense. It is your responsibility to make sure that your activities comply with applicable laws, including copyright.

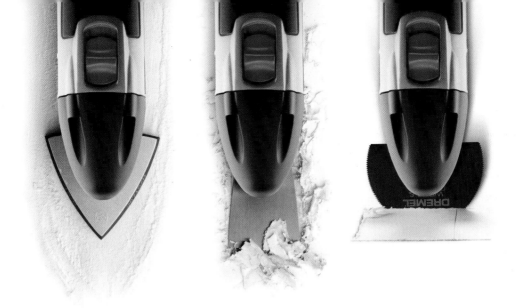

From the company that invented versatility, here's some more.

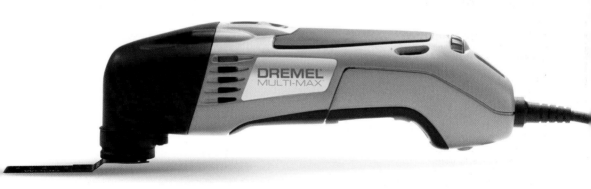

Cut	Remove	Grind	Scrape	Sand
wood, drywall, metal	grout, caulk, paint	thinset, cement, mortar	adhesive, paint, vinyl	wood, metal, chipboard

The all-new Multi-Max™ Oscillating Tool System.

Its fast side-to-side motion and compact design make it the perfect tool for even your most demanding remodeling projects. From cutting a door jamb to removing grout—and every job in between. It's exactly the kind of versatility you've come to expect from a Dremel tool. Call 1-800-437-3635 today for a free DVD. Or visit dremel.com to view project videos.

DREMEL®
MULTI-MAX™
Repair. Remodel. Restore.™

Contributors

Adam Koford (Cover and Special Section illustrations) is "a father, an Eagle Scout, pretty tall, and a cartoonist." His first drawing of a butterfly didn't impress his mom: "She was pretty sure I had traced it, so she had me redo it." He works for Disney Interactive Studios in Salt Lake City as a story artist, and has a daily webcomic called Laugh-Out-Loud Cats and a great recipe for grapefruit cake (ask him about it — we did).

Mike Gould (*Lunchbox Laser Shows*) is a longtime builder of devices, Macintosh expert, webmaster, commercial photographer, technical writer, camper, and hunter of morels. He bought his first laser in 1972 and worked with the Martian Entropy Band as a bass player and light show guy with his buddy Wayne Gillis. "Wayne and I are still building gizmos together 37 years later," he says. Mike lives in Ann Arbor, Mich., with "one very patient wife," and is currently working on Illuminatus 2.1: "Having mastered lunchboxes, we are now building lasers into antique projectors and stage lights as steampunk display artifacts."

Jason Hornick (*10-Rocket Mega-Launcher* photography) is an editorial/commercial photographer based in Washington, D.C. After starting off as a staff photographer at newspapers in the D.C. metro area, he moved to full-time freelance status a year ago. Over the last six years, photography has allowed him to explore many places across the United States and the world. He's photographed the best and the worst of life and had the fortune of meeting the most interesting people following their passion in life. When he's not making photographs, Jason enjoys getting into the outdoors and mountain biking, hiking, and camping.

Serendipity played a role in the career of **Patrick Keeling** (*Van Leeuwenhoek Microscope*). In college, he confused molecular biology (the study of DNA) with microbiology (the study of microbial life) and signed up for a bunch of microbiology courses. "It was a happy accident, because it was all pretty interesting and I now study the molecular biology of microbes." The Vancouver, B.C., father of two is also a woodworker and outdoorsman. What's exciting him lately? Biological "barcoding" has him thinking "about whether (or not) our concept of 'species' works for microbial cells like it does for plants and animals, and just how we can think about diversity in the microbial world."

James Provost's (*The Maker's Bill of Rights* and *Marble Adding Machine* illustrations) childhood was spent playing with computers, drawing, and taking things apart to learn how they worked. His first job was at Subway. His last job was drawing subways. Since then he's been a freelance technical illustrator who spends his time playing with computers, drawing, and taking things apart to learn how they work.

Dan Rasmussen (*Auto-Phenakistoscope*) is an avid collector and fixer of vintage technology and an IT specialist for IBM who lives in Groton, Mass., with his wife, three kids, a dog, a very old cat, a hamster, and dozens of soon-to-be frogs. Why vintage technology? Understanding how earlier engineers "solved problems under a different set of circumstances/constraints is not only fascinating but can provide insights into how to solve new problems." Plus, he says, "My favorite thing to do is to make things work again. There is nothing like hearing a radio come back to life after 50+ years of silence." Next up? Teaching old radios how to play only old radio shows.

Watch, Learn, Do

Observing my father make things when I was young instilled confidence in my own abilities.

Growing up, I was given specific advantages as a maker. My father, the painter Lee Savage, was a living example of a committed maker: painter, animator, illustrator, director. My primary memory of him, growing up, is of him painting every day for hours in his studio out back, and living in a house populated with art made by close friends.

When, for my sixth or seventh birthday, I wanted a race car for my teddy bear, Gus, he made one out of fiberglass for me. He built a succession of decks behind our house, built the addition on his studio, and fixed the chairs he broke leaning back too far. I watched it all.

His studio was a laboratory of excellent primary building materials: mat board, Rapidograph pens, acetate (which he used to paint cels for animated spots he did for *Sesame Street*), armature wire, and masking tape. I was never turned away for want of an art material.

By the age of 12, I had permission to use my father's charge account at the local hardware store. At 18, when I moved into Manhattan, he let me charge art materials on his account at the art store. I was stoked. When I got ambitious, and asked for things like 20 sheets of corrugated cardboard, the answer was always yes. I never abused this privilege. It honestly never occurred to me.

I don't imagine that my father set out to create an artist. But I'm pretty sure he figured that everyone has a responsibility to learn how the world went together, and much of that learning is simply paying attention.

My twin boys, Addison and Reilly, are now approaching 11. I don't spoil them, but I do want them to be makers. So I don't end up building a ton of stuff *for* them.

But: I've bought models for them, and shown them how to put them together. I'm always using one or the other as an assistant for my projects, whether fixing something in the house or my car, or putting together a piece of a costume. They've spent the day at work with me many times. They

like to watch, and love to help, and already I can see the fruits of these efforts starting to bloom.

I can see both boys learning to put something they want to make into their heads. I can see them trying, failing, succeeding, and trying again to get that thing made.

They're both showing a facility for music. I'm encouraging the spit out of this. The ability to enjoy doing a thing excellently, the ability to enjoy the *work* involved — to know that trial and error and even failure may lie ahead, but that they aren't enough to inhibit your forward progress — this is what I hope to teach them.

Part of gaining the courage to plug ahead with anything is acquiring the confidence that you're going to be able to understand what's going to go on. The more things you build and make, the more things you take apart and break, the more you understand many of the critical workings of the world. The more you pay attention, the more that attention pays you back.

Adam Savage is an American industrial designer, special effects designer/fabricator, actor, educator, and co-host of the Discovery Channel television series *MythBusters*. He lives in the San Francisco Bay Area with his wife, Julie, and two sons.

Illustration by Adam Koford; photograph by Cody Pickens

Young makers, DTV praises, a melted museum, and a CRAFTer's lament.

✉ MAKE, Volume 19, is by far my favorite issue. The only thing I like more than a week in the shed with my tools and MAKE magazine is a week in the shed with tools, MAKE, and my son! Not only were there lots of projects for him *and* me, but he especially enjoyed the photos of young people who enjoy building too. (Dad says: good examples.) My wife and I hope this continues to be a regular feature. Great job! Thanks for remembering the younger builders, like my son, Marty.

—*Matthew and Marty Ruane, Richland, Wash.*

✉ Thank you for an incredible magazine. I love it! In fact, we use MAKE in our museum's Science Center for inspiration and to design experiments and exhibits for our visitors.

At the Science Center we have opened an exciting and very special exhibition about global climate change. Visitors have to put on rubber boots as the exhibition floor is covered with 10cm of water to illustrate the effect of increasing sea level.

Enormous melting ice cubes symbolize the melting of the Arctic ice cap. The visitors use remote controlled boats to visit kiosks representing different geographical places around the world. By entering one of the kiosk's harbors with the boat, the visitor can start a short movie featuring a climate witness. The exhibition is very inventive and exciting, just like MAKE.

—*Jon Haavie, Oslo, Norway*
Norsk Teknisk Museum, teskniskmuseum.no

✉ I really enjoyed the *Make:* television program on how to make a DTV antenna [Episode 4 at makezine.tv] , and I tried it. We were having problems getting TV reception on our old antenna. The signal strength was only 18%. After building the DTV antenna, all channels came in sharp and clear with a signal strength of 65%.

I only made a few changes on the one I built. Where the coat hangers crossed, I used heat-shrink tubing instead of electrical tape, and for a better connection, I used 2 washers where the transformer connects, sandwiching the connections between them. Thanks for the good and useful project.

—*Len Hart, Niangua, Mo.*

✉ I am one of the sad saps who received MAKE as the consolation prize when CRAFT folded. This was not a chance for me to experience something new, since I was already a MAKE subscriber. I read that MAKE had planned on expanding to satisfy what CRAFT was doing, and for the last few issues since the end of CRAFT, I have felt satisfied. Now, with Volume 19, I feel like you have shaken off all former CRAFT subscribers to return to a more digital/tech-heavy focus. As far as some constructive criticism, here are things that I'd like to see:

• More projects that lean towards mechanical skills to balance out the digital.
• More projects about food/alcohol/beverage science and production.

Oooorrr, you could scrap that and just bring back CRAFT. I would still subscribe to both.

—*Emily Armstrong, Troy, N.Y.*

EDITOR'S NOTE: Emily, we're glad you're sticking with us and we remain committed to featuring all kinds of making in MAKE. Naturally Volume 19, the Robots issue, had lots of technology; we hope you'll enjoy this new issue focused on projects for Kids of All Ages. And don't forget about our content-rich website, craftzine.com, updated daily.

MAKE AMENDS

In Volume 19, page 79, "My Robot, Makey," the 0.1µF capacitor is Jameco part #151116, not #15229. Thanks to reader Joseph E. Mayer for catching the error.

Also in Volume 19, we omitted Daniel Klaussen, who tested the instructions for the Speed Vest project, (page 100). Thanks, Daniel, for all your help!

Learn by Making

For more than seven years, Tom Zimmerman has volunteered in San Jose, Calif., schools, engaging students in hands-on activities and teaching science and technology. This summer, Zimmerman was recognized as the first-ever California Volunteer of the Year by Governor Arnold Schwarzenegger and First Lady Maria Shriver. He called it a "fairytale day."

Zimmerman is an IBM research scientist and a frequent contributor to MAKE — his Hydrogen-Oxygen Rocket project is featured in this issue (*page 90*). Much of what he's written for the magazine originated as projects he developed for his students: an electronic drum kit, a mini Mars rover, and a digital microscope.

He created an Extreme Science after-school program to provide 60 Latino high school students with hands-on experience in science, technology, engineering, and math (STEM). And he's particularly proud of his efforts to introduce girls to power tools. He also runs a summer camp, which this year featured workshops on building wind turbines and a geodesic dome. "I'm happy to have the opportunity to share the joy of designing, building, and teaching," he enthuses.

In Boston, Ed Baafi runs the Learn 2 Teach, Teach 2 Learn (L2T) program during spring and summer at MIT and at the SouthEnd Technology Center and its FabLab. The idea behind L2T is that the best way to demonstrate that you've learned something is to turn around and teach it to others.

"We pay high school students to learn, build, and teach at over a dozen community centers," says Baafi. One group developed a solar device charger, and another student, Mark Williams, has been perfecting his electric violin, which we blogged on makezine.com, to his great amazement.

A seventh-grade teacher explains why she came to Maker Faire this year:

I try to incorporate some hands-on activities and labs into the classroom. I am still dissatisfied with the learning environment I am able to provide to my kids. The "holy grail" for me is to facilitate communities of independent learners, engaged in projects, assignments, discussions, etc. that motivate and challenge them. To this end, I'd like to make more stuff, and to have my students be makers.

Our communities are made up of makers like Zimmerman and Baafi as well as teachers and parents who see the importance of helping kids become makers.

But it's become clear that making is missing from schools and from the lives of even the best students. "I have had freshman engineering students who have never used power tools," says AnnMarie P. Thomas, an engineering professor at the University of St. Thomas in St. Paul, Minn.

How can we create more opportunities for kids to make things? How do we create spaces inside schools or out in the community that support self-directed, hands-on projects?

Making is a way to engage kids in learning. It's not work but a form of play. "There is a kind of magic in play," writes Stuart Brown in his new book, *Play*. "It's paradoxical that a little bit of 'nonproductive' activity can make one enormously more productive and invigorated in other aspects of life."

Making is a way to enjoy trying to do new things (and often failing repeatedly) while learning more than any written test can measure. "Allowing children to build with real tools," says Thomas, "gives them confidence and a skill set that they can build on for years to come."

This magazine will do its best to advocate for the role of making in education. Makers themselves are an untapped resource for schools, especially as mentors. I know many makers who are exploring ways to share what they know and love with kids of all ages. Makers bring more than knowledge and experience — they bring endless enthusiasm, which they easily pass on.

If you're interested in making and education, get involved in your own community. Join me and others at Make: Education (makered.makezine.com) to share ideas, stories, and techniques for helping more kids learn by making.

Dale Dougherty is the editor and publisher of MAKE magazine.

Shortcut to Omniscience

Talk to anyone who's tried to edit a controversial Wikipedia article and chances are you'll be treated to an earful of complaint. Wikipedia's legion of committed editors tirelessly revert the changes wrought by newbies within minutes, often with flat declarations of the ineligibility of the new material for the online encyclopedia. This proves frustrating for skilled contributors who want to correct errors.

To understand what's going on here, you need to appreciate what makes Wikipedia possible. Prior to Wikipedia's launch in 2001, the consensus was that millions of amateurs would never be able to write an encyclopedia together. Encyclopedias require two things: control and expertise, and the Wikipedia project made very little room for either.

The thing is, expertise and control are expensive. They're the kind of thing a publisher can raise and charge money for, but they're not readily available to informal groups of amateurs. Wikipedia's success hinged on figuring out how to get around this handicap, and the solution was so clever — and frustrating — that most people miss it altogether.

Here's the thing about expertise: it's hard to define. It may be possible for a small group of relatively homogenous people to agree on who is and isn't an expert, but getting millions of people to do so is practically impossible. The *Encyclopædia Britannica* uses a learned editorial board to decide who will write its entries and who will review them.

Wikipedia turns this on its head by saying, essentially, "Anyone can write our entries, but those entries should consist of material cited from reliable sources." While the *Britannica* says, "These facts are true," Wikipedia says, "It is true that these facts were reported by these sources." The *Britannica* contains facts. Wikipedia contains facts about facts.

And this is Wikipedia's secret weapon and its greatest weakness. The debate over which sources are notable is a lot more manageable than the debate about which facts are true (though the former is nevertheless difficult and it consumes many Wikipedian-hours). Moving to a tractable debate about sources makes it possible for millions of people to collaborate on writing the encyclopedia.

But this shortcut also creates endless frustration.

> The *Encyclopædia Britannica* contains facts. Wikipedia contains facts about facts. This is Wikipedia's secret weapon and its greatest weakness.

You might not be able to correct the Wikipedia date for a famous battle that you're a worldwide expert on, by asserting that you know it is incorrect. But give an interview to *The New York Times* about how screwed-up Wikipedia is and cite the true date, and you can go back to the erroneous entry and correct it without argument, by citing the fact as published in *The New York Times*.

This seems completely backward and absurd at first, but remember: Wikipedia is a collection of facts about facts. It's incredibly hard for the whole world of Wikipedians to look up your credentials and decide that you know what you're talking about; however, it's simple for the editorial world to look at *The New York Times* and see what they've reported. And since there's consensus that the *Times* is a notable source (notwithstanding Jason Blair and other scandals), the edit can now stand.

Being a maker in a networked society often involves moving processes from giant factories into humble garage workshops, thanks to cheap, flexible tools, readily available materials, and easy knowledge-sharing. But remember: the way you make depends as much on how you're organized as on what you want to make.

When you run up against the limits of what you can do in your garage, ask yourself: Is the way I'm doing this inherent to what I'm doing? Have I imported some 20th-century organizational style that's holding me back?

You never know: there may just be more than one way to skin an encyclopedia.

Cory Doctorow lives in London, writes science fiction novels, co-edits Boing Boing, and fights for digital freedom.

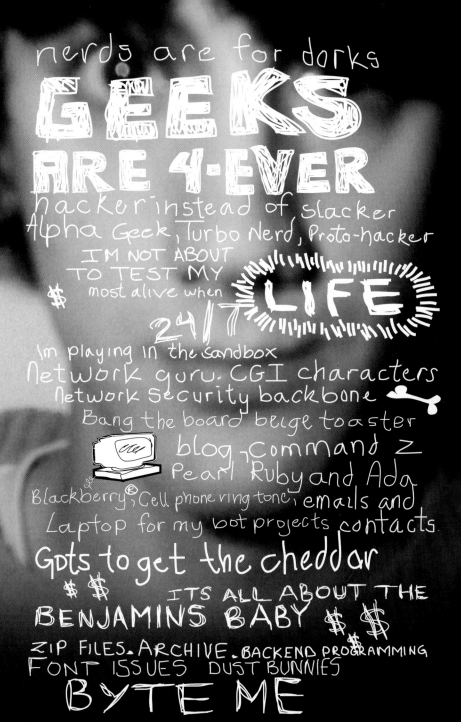

nerds are for dorks

GEEKS
ARE 4-EVER

hacker instead of slacker
Alpha Geek, Turbo Nerd, Proto-hacker
I'M NOT ABOUT
TO TEST MY
$ most alive when
24/7 **LIFE**

Im playing in the sandbox
Network guru. CGI characters
Network security backbone
Bang the board beige toaster
blog command z
Pearl Ruby and Ada
Blackberry® Cell phone ring tone, emails and
Laptop for my bot projects contacts.

Gots to get the cheddar
$ $ ITS ALL ABOUT THE
BENJAMINS BABY $ $
$
ZIP FILES. ARCHIVE. BACKEND PROGRAMMING
FONT ISSUES DUST BUNNIES
BYTE ME

Baby Hacking

Having a child to build things for is the maker's dream, and the maker's curse. I had a son seven months ago. It is every bit as wonderful, and every bit as exhausting, as everyone will tell you. As a hacker, a maker, and a builder, I find that it's definitely a humbling experience. This is, quite literally, my life's work — in the strict biological sense.

It's also humbling because he is so much cooler than anything else I'll ever make (or half make). So far he's done little more than transform from a strictly input/output device to an interactive robot; all the same, watching his operating system boot up makes any code I've ever written look trivial.

In observing his every movement, I can only be jealous of the evolutionary mechanism design-and-optimization that is his every muscle, digit, earlobe, and nostril. Because this is the pinnacle of my making, I'm rattled, but I'm also inspired, because he just set the bar so much higher.

If you suffer from inventor-itis, as I do, the first thing you notice about the baby world is that everything is broken. All the products on the shelf are pathetic. They're toxic and poorly made. The bed options are hopeless. Those clumsy, bulky, awkward, ugly things called strollers are terrible. Car safety seats? I wouldn't trust them to a high-school egg-drop competition. Enter the maker mother and father: time to get a fixin' on it.

But there's a downside. Time vanishes when you have a kid. Those moments of peace, contemplation, and low-level hand-eye tasks you used to have alone at the workbench? Obliterated. There is now a screaming, giggling attention magnet that's the cutest thing in the world. You haven't slept in weeks.

Hacksawing with the kid in the sling on my chest? I don't think so (or at least my wife has trained me not to think so!). Actually getting things built on time? That's a whole other story. My wife axed my goal of building the ultimate stroller, when the baby arrived and there was still no stroller. (Perhaps for the next kid, I wistfully think.) And my folding origami toy box concept will likely remain a dream.

So far I've managed to build a co-sleeper (a mini bed that attaches to our bed) with the help of a friend. Most are hideous and cheaply made — there was one in a design magazine somewhere that we liked, but it was only a concept! So I made it real, a bamboo and aluminum masterpiece.

Our cargo tricycle wasn't designed with an infant sunshade, so my father and I fixed it. My wife supplied beautiful Marimekko fabric so the outcome wouldn't be embarrassing at the new mothers' group, and the result is a magnificent, aerodynamic sunshade made by two engineers with PhDs.

I find myself working on children's toys out of compulsion. I have redefined the pinnacle of invention as "the next Lego." It's crazy and arrogant to believe that you could do it, but that won't stop me from trying. I'm sure you know what I mean. I want my child to have amazing experiences, to grow up in a world of objects that are beautifully designed and thoughtfully made. Toys that are significant and memorable, not disposable.

Perhaps that's the hidden desire I'm talking about. It's not that my own childhood wasn't magical (it was; my father used the best tools of his day to make me ride-on wooden horses, pedal-powered cars, and knitted 8-foot-high unicorns).

My love, and passion, and exuberance, and creative desire are overflowing, and every sheet of plywood or steel that I look at is some incredible object that I can make for him. Inflatable safety barriers for car seats? Easy. I'll do it for him. Custom stuffed-animal-creation software? I'll do it for him. An algorithm-based paper airplane generator? I'll code it for him.

And that baby stroller? It was going to be made from aircraft-grade aluminum poles, cast-zinc universal connection pieces so it could be reconfigurable, and Abec 11 bearings with large-diameter rollerblade wheels (for low rolling resistance). It would fold into something smaller than a Swiss Army knife yet be sturdy as a tank.

All the drawings are still in my head. But the baby needs feeding, my wife needs a break, I need some sleep, and the stroller needs one of you non-parent makers to make it for us. Unfortunately, until you have a child, you probably won't quite understand why.

Saul Griffith is a new father and an entrepreneur.
otherlab.com

Get a jump start on your holiday shopping —
Mail the Make gift form today!

makezine.com/subscribe
for faster service, subscribe online

NO POSTAGE
NECESSARY
IF MAILED
IN THE
UNITED STATES

BUSINESS REPLY MAIL
FIRST-CLASS MAIL   PERMIT NO 865   NORTH HOLLYWOOD CA

POSTAGE WILL BE PAID BY ADDRESSEE

Make:
technology on your time

PO BOX 17046
NORTH HOLLYWOOD CA 91615-9588

A Major Model Village

Why are models so fascinating? Psychologists haven't spilled much ink over the question, but just about everyone visiting Bekonscot, the world's oldest model village, is quickly spellbound.

The whole thing began by accident in the 1920s when **Roland Callingham**'s wife issued an ultimatum: either the model railway in the house goes, or I do. It appears that Callingham didn't want to part with either, so he bought four acres of meadows next to his house in Buckinghamshire, England, near London, and proceeded to build an outdoor model railway, along with a model world to go with it.

With the help of model rail manufacturer Bassett-Lowke (still in existence), Callingham's ad hoc team built a robust Gauge 1 line at ¹⁄₃₂ scale. Amazingly, one of the locomotives from 1929 is still running — still doing 2,000 (real, not model) miles a year.

The initial rail line skirted the new swimming pool that Callingham had dug for his tennis guests to cool off in. The earth from the pool excavation formed mounds on which the first model houses were built to complement the growing train line.

The houses were built to a different scale of 1 inch to 1 foot, now standard for many model villages.

Bekonscot opened to the public in 1929 and was an immediate success, even receiving several royal visits. It has expanded to fill nearly two acres with 400 yards of railways, six villages and seven stations, hundreds of buildings, and thousands of figurines — all made on-site by modelers and engineers, many of whom work there for decades.

Bekonscot is a charming snapshot of Britain in the 1930s, but it also encapsulates much of the history of industrial civilization. It has a coal mine, an oil refinery, an airport, a racecourse, a hospital, a farm, a funfair (carnival), churches, and schools, as well as lakes, a fishing harbor, and a yacht marina. It is a sensual and intellectual *tour de force* that's attracted nearly 15 million visitors, given more than £5 million ($8 million) to charity (in today's money), and been the inspiration for countless other model villages around the world.   —*Julian Darley*

>> **Bekonscot Model Village:** bekonscot.com

WALL-E World

Before the Disney/Pixar film *WALL-E* premiered in 2008, replicas of its robot star were already showing up on the internet. That's because at least one replica builders' group had a head start.

Scot Washburn explains: "I found an early trailer for *WALL-E* and posted it to the R2-D2 Builders Club around the end of September 2007. There was such a positive response that ... on October 8th, I created the WALL-E Builders Club."

The club has grown to nearly 700 passionate bot replica builders who mainly exchange information online, but sometimes meet at nerdy conventions.

So, what inspires grown men (almost exclusively) to spend hundreds of hours and thousands of dollars to replicate a cartoon robot?

"He's a very sympathetic character," says club member Guy Vardaman, a web developer from Burbank, Calif. "He's innocent-looking — yeah, those large eyes. But he also looks somewhat rugged, with his tank-tread drive and all of those scrapes and dents on his body. He looks realistic, too; like you can believe he'd actually work."

Maybe that's what launched a thousand WALL-Es: he looks easy to make, at least as a static replica. Articulated WALL-Es, with a working drive, sound effects, and radio control, are harder to come by.

"My WALL-E will be radio controlled, so he'll move around on his treads," says William Miyamoto, 42, a stay-at-home dad and actor from Los Angeles. "I plan on articulating his head and arms. He will also have a sound system so he'll say things and play sounds on command."

The club is collaborating to design a track drive that members can replicate. Members point out a myriad of benefits to working in a group, such as the pooling of talent, expertise, and purchasing power, plus the trading and sharing of parts.

But there are also the more human aspects. As one member put it: "When you're building something that takes months or years to finish, you can run out of steam. The encouragement of the group can make all the difference."   —*Gareth Branwyn*

≫ **WALL-E Builders Club:** makezine.com/go/wall-e

Photograph by Cory Pacion

Bug Warriors

A new breed of bionic air and ground forces — off-road tarantulas, scorpion tanks, butterfly copters, beetle jets, and more — rolls, crawls, and flies across the battlefield in a cyborg army designed and built by artist **Dean Christ**.

At least we think that's who he is.

"I can only reveal that my name is Dean Christ and I'm currently based in Sydney, Australia, after living in Japan for several years," he says. "That's all you really need to know."

OK. So where did the cryptic creator come up with his nature-hacked troop concept? Not in the pages of science fiction, but by reading a blog post a year ago about a U.S. Defense Advanced Research Projects Agency (DARPA)-funded study aiming to remotely control moth flight using integrated electronics. A couple months later, he stumbled across museum-grade scorpion and spider specimens at a local weekend market.

"It clicked," he says. "I like finding inspiration in concepts or ideas that at first seem absurdist or unbelievable but are grounded in some sort of reality. I guess the saying 'fact is stranger than fiction' sums it up."

Next, he purchased his own specimens from suppliers around the world and added scale-model military parts to suit each one. Scorpions, with their armored shells, seemed to fit a tank design; butterflies are light and hover like helicopters; beetles have sleek, streamlined bodies with hidden wings — the perfect fighter jets.

To date, the cyborg army consists of 30 creatures wielding an arsenal of technologies like stealth, smart bombs, rockets, and Stinger missiles. Christ hopes to exhibit in the coming year.

The bug battalion represents what Christ fears could become real if projects like DARPA's are taken to the extreme. To him, the cyborgs could be the next step in biological warfare, and the next arms race may involve pincers and antennae along with grenades and bombs.

—*Megan Mansell Williams*

≫ **Biomechanical Bug Army:** cyborganimals.com

Treespotting

In recent years, the abandoned lots and burned-out houses of downtown Detroit have given way to urban prairie. Weeds, vines, even trees poke out of broken windows, and one plant in particular, *Ailanthus altissima* — aka Tree-of-Heaven or "ghetto palm" — runs rampant. Able to grow a ravenous 5 feet a year, flourish in toxic soil, power through walls, and resist the occasional bullet, the exotic tree has become synonymous with urban blight.

But where some see an invasive pest, **Mitch Cope** sees opportunity. In 2005, Cope along with fellow artists **Ingo Vetter** and **Annette Weisser** founded the Detroit Tree of Heaven Woodshop, a collective of artists, arborists, and woodworkers that transforms the ghetto palm into lumber.

In a ritual they call "treespotting," Cope and crew scout the streets of the Motor City for prime specimens. Once they've got their tree, arborist **Kevin Bingham** cuts it down and then they call up Last Chance Logs to Lumber, a small urban milling company that processes the wood on-site with a portable band saw.

Curing can be tricky. The tree's large pores release a whopping 5 gallons of water a day during drying, which makes the wood susceptible to warping. A trial run with a solar kiln dehydrated the wood too quickly and made it crack. So far, the most effective technique has been drying it slowly outdoors under shelter. Cope admits experienced woodworkers have been stumped and surprised by the results.

The Woodshop turns out furniture and sculpture including a set of sleek, minimal benches for an exhibition at the Museum of Contemporary Art Detroit. And although the project is essentially a design experiment, not a moneymaking venture, Cope believes that given the tree's overabundance in urban and rural areas, an enterprising soul could turn their model into a viable business. Who said Detroit was out of ideas? —*Jeanne Storck*

≫ **The Woodshop:** treeofheavenwoodshop.com

The Detroit Tree of Heaven Woodshop is taking part in Heartland, an exhibition of Midwestern art and ingenuity at the Smart Museum in Chicago. Oct. 1, 2009 — Jan. 17, 2010.

Photography courtesy of Tree of Heaven Woodshop

Hit Me with Your Stretch Shot

"DANGER! This rubber band gun ... is capable of inflicting 1st and 2nd degree welts, contusions, and bruises on the supple flesh of its targets," reads the packaging on the Rubber Bandit. "For the lady or gentleman that knows better than to grow up."

Its creator, 21-year-old **Andy Mangold**, never did learn. He built his first rubber band gun in middle school out in the garage of his Pennsylvania home. "I made an arsenal of them that included a Gatling that could fire 40 rubber bands as fast as you could turn the crank. You had to reload for like 20 minutes, but it was worth it for that moment of glory."

"When he shot the Gatling, everyone took cover," says Andy's mom, Linda. Opposed to buying toy guns, she couldn't really complain once he started making his own. "The way I looked at it was at least he had to use creativity."

Mangold shot one too many bands at his sister, though, and was forced into early retirement.

But Mangold's boyhood playtime would later become his silver bullet. Now a respectable intern at Shaw Jelveh Design in Baltimore and a junior at the Maryland Institute College of Art (MICA) majoring in graphic design with a concentration in book arts, Mangold knew just what to do when a class assignment called on students to design, manufacture, and package a toy from scratch. His sure-fire secret weapon: the good ol' rubber band gun.

After building a plywood dummy, he cut and carved the final product from exotic curly maple and bloodwood. Without using springs, instead just the forces of the rubber bands and wooden gears, his latest gun holds five bands and fires one at a time without having to reload, and it comes with interchangeable barrels (longer ones shoot harder). The packaging incorporates bookboard and newsprint for a vintage, faded look.

His sister knows he's back at it. "She saw it," Mangold says. "It brought back some nightmares. She has a little post-traumatic stress."

—*Megan Mansell Williams*

≫ **The Rubber Bandit:** andymangold.com/the-rubber-bandit

Oilcan Canopy

What happens when a thousand oilcans decide to fly? That's what **Sanjeev Shankar** and the residents of Rajokri, India, recently tried to find out in a grand project reusing old oilcans. Called *Jugaad* after the Hindi term for jerry-rigging or MacGyvering, Shankar's creation uses a ubiquitous piece of trash to provide respite from the afternoon sun.

Constructed for the 48°C Public.Art.Ecology festival in New Delhi in December 2008, *Jugaad* is a suspended shade pavilion made from 692 discarded oil cans and 945 oil can covers, spanning 750 square feet. The covers were hand-painted with gulal, a local pigment, and stitched together using thin metal wires to create a pixelated, pink metal fabric. Halogen light fixtures were placed in the existing openings to provide illumination at night.

Shankar hopes to raise the profile of sustainability and to redefine repurposing. "*Jugaad* takes recycling and reuse beyond a simply utilitarian measure, into an exciting world of architecture and design possibilities," he says.

Shankar is an Indian artist, architect, and designer who says he likes to merge traditional crafts-based knowledge with more contemporary and emerging cultural and technological trends. An alumnus of the Indian Institute of Technology, his work has been seen all over the globe, from Brussels to New York to Bombay.

The *Jugaad* project was also about community participation. The people of Rajokri, an urban village at the edge of New Delhi, helped to collect the cans and build the pavilion. Although the villagers were initially hesitant to modify discarded oilcans, Shankar says that engaging them in a participatory design approach helped overcome this. Resistance gave way to enthusiasm, and a cottage industry sprung up around reusing the cans.

"The most rewarding part of *Jugaad*," he explains, "was witnessing the combined human spirit of creativity, improvisation, and celebration in a culture of scarcity and survival."

—*Bruce Stewart*

≫ **Oilcans in Flight:** sanjeevshankar.com

Pedal Smoothie

As a former competitive swimmer who could do 11 pool miles a day, **David Butcher** knew the human body could crank out a couple hundred watts of work for sustained periods. What if he could turn muscle power into electricity?

Having witnessed the 1969 Santa Barbara oil spill, Butcher is a dedicated environmentalist. He wanted to get off the grid, but he found solar and wind projects unfeasible. "That left *me*," he says.

So he built the Pedal Powered Prime Mover (PPPM): a souped-up stationary bike generator that boosts efficiency with a large particleboard flywheel that smooths out the torque peaks of pedaling.

Its generator is a sealed ball-bearing motor from a Razor scooter, driven by a BMX crankset. Completing the setup are a 1,000-watt inverter, Watts Up power meter, and Maxwell Technologies 15-volt, 58-farad ultracapacitor module to store power quickly and dump it on demand.

Butcher, 55, has clocked his peak output at 265 watts (⅓ horsepower), but reckons a younger rider could double that. At his home in San Jose, Calif., he

uses the PPPM to power laptops, TVs, a Roomba, even a front-loading washer (with the help of a battery boost to surmount the spin cycles).

In his video on *Make:* television, he cranks up enough juice to run a 500-watt blender and puree a liter of smoothie in under 60 seconds. He's cut his energy bills, dropped 40 pounds, and gotten super fit.

"The longest I ever rode was at Maker Faire 2007, from 9 to 6," Butcher recalls excitedly. "I also watched the entire *Battlestar Galactica* finale, powering it myself. That was a big one for me!"

Butcher sells detailed DIY plans for the PPPM and promotes it for everyday and emergency power, and for use in remote villages. But he says the most efficient use of pedal power isn't electricity — it's direct mechanical connection to turn a pump, fan, or the like. As evidence, he's also built a pedal-powered pickup truck and a pedal-powered canoe.

—*Keith Hammond*

» **Pedal Powered Generator:** makezine.com/go/pedgen

Make: television Episode 110: makezine.tv/episodes

Photograph by Ed Troxell

Makezine.com Rules the DIY Web

MAKE is one of the planet's most popular DIY blogs and RSS feeds, with bloggers like Phil Torrone and Gareth Branwyn, our popular Weekend Projects videos, MAKE and CRAFT video and PDF podcasts, fun contests, and thousands of projects from around the world. Bookmark us, or subscribe via RSS, at blog.makezine.com. Here are a few of our favorite online offerings from recent months:

M: In his **DIY Cymatics video**, Collin Cunningham shows how to make a "non-Newtonian fluid" (cornstarch and water) come alive in bizarre, writhing patterns, using sound. See it to believe it, at makezine.com/go/cymatics. And don't miss Collin's **Make: Presents introductory electronics videos**, from the humble resistor to the mighty microchip, at makezine.com/go/makepresents.

M: For an awesome **Weekend Project**, build the **Night Lighter 36 spud gun** from MAKE, Volume 03. Sparked by a stun gun, it launches potatoes 200 yards and lights up with see-through action! Kipkay shows you how at makezine.com/go/nightlighter.

M: Digital TV is here and people are flocking to make their own **DTV antennas** as demonstrated by John Park on **Make: television**. Watch Episode 4 at makezine.com/go/dtv or grab PDF instructions at makezine.com/go/dtvpdf. Even *Consumer Reports* recommended us!

M: If you've upgraded to HDTV, what do you do with your old set? Becky Stern tracks down **DIY ideas for surplus TVs** in our weekly **Ask MAKE** column, at makezine.com/go/surplustv. Email your questions to becky@makezine.com or Twitter @make.

M: Got a Maker's Notebook for your ideas? Color your fevered scribblings with the easy **Thermochromic Maker's Notebook hack**, at makezine.com/go/thermochromic.

M: MAKE and CRAFT's sharp young interns have a great job: make stuff from the magazine and test it out. In **Intern's Corner**, they blog weekly about the projects they're building and the trouble they get into, at blog.makezine.com/archive/interns_corner and blog.craftzine.com/archive/interns_corner.

Collin's Lab Notes: DIY Cymatics

Weekend Project: Stun Gun Potato Cannon

M: Check out all the winning projects from MAKE's killer **DIY Halloween contests**, and our summertime **MAKEcation contests**, where families learned how to solder, hack their drink coolers, and build trebuchets. Go to makezine.com/halloweencontest and makezine.com/go/makecations.

M: Citizen scientists, backyard explorers, teachers, and budding biohackers, our new DIY science destination **Make: Science Room** was made just for you. With its many resources, use it as your **DIY science classroom**, virtual laboratory, and a place to share your projects, hacks, and lab tips with other amateur scientists. Hosted by author Robert Bruce Thompson. Go to blog.makezine.com/science_room.

Doing Science With a Digital Scanner

The transition from film to digital cameras has made a huge impact on how amateur scientists can save, analyze, catalog, and publish their imagery. It's safe to say that digital cameras and personal computers are among the most important tools in the amateur scientist's kit.

When the subject is two-dimensional, flatbed digital scanners can also play a major role in imaging science. They are ideal for making high-resolution images of leaves, dragonflies, butterflies, tree ring sections, soil samples, and many other subjects.

Advantages and Disadvantages of Scanning

Virtually shadow-free lighting is a key advantage of a digital scanner, for the scanner provides its own light source. Another advantage is that scanners don't suffer from the distortions caused by camera lenses. Scanners are relatively inexpensive, and they can be used for many applications beyond the scientific roles described here.

Besides their two-dimension limitation, a major drawback of scanners is that objects being scanned must fit within the scanner's image plane. Scanners are also much larger than digital cameras.

Background Color

Most objects that I've scanned looked best with a white background. Because my scanner (HP Scanjet 3970) has a 35mm slide scanner slot built into its lid, it's necessary to cover the object being scanned with an uninterrupted background. Two or three sheets of white 20-pound paper work well.

Light-colored objects are not easily visible when scanned against the white background of a typical scanner lid. To provide contrast with light-colored objects, place black construction paper over the object being scanned before closing the lid. You can use various colors for special effects.

How to Scan Dragonflies

For years I have photographed dragonflies and damselflies by quietly sneaking behind them. With practice, it's possible to get within a few inches of some species. While real-world images like these are important, entomologists Forrest Mitchell and James L. Lasswell of the Texas AgriLife Research and Extension Center at Stephenville, Texas, have used digital scanning to create an impressive library of top and side views of many dragonflies and damselflies. You can see these images online at Digital Dragonflies (dragonflies.org).

The images at Digital Dragonflies were made by cooling each specimen in a refrigerator to keep it still while it was being scanned. The insects were then placed backside down on the scanner's glass bed. They were protected from being crushed by the scanner's lid by placing them inside a 10cm×12cm rectangle cut in a mouse pad (see their website for details). The mouse pad approach is probably best, but I've found that ordinary corrugated fiberboard will also work.

Based on my experience scanning dragonflies (Figure A), it would seem that butterflies and moths could also be scanned. A severe drought has slashed the butterfly population in Central Texas, and I'll try this method as soon as the butterflies return.

Fossils and Artifacts

Fossils and artifacts having a flat surface are easily scanned. For example, I've scanned the fossil of a trilobite with good results (Figure B). Even ripple marks in the mud (now shale) on which the creature was resting were captured.

Years ago I found a flint artifact in the gravel bottom of the creek that borders our land. The artifact is a flint scraper that is very flat on both sides, and it was easily scanned against a white background (Figure C). Flint arrowheads and spear points can also be scanned.

Plants

The leaves of many kinds of plants are easily scanned against a white background (Figure D). The main limitation of scanning leaves is the size restriction posed by the scanner's bed.

Fig. A: This living dragonfly was scanned by placing it inside an opening cut in a sheet of corrugated board. **Figs. B and C:** Many fossils and flint implements are flat enough to permit scanning, like this trilobite and scraper. **Fig. D:** Be sure to clean the glass surface of your scanner after scanning plant leaves. **Fig. E:** Feathers are easily scanned. **Fig. F:** This piece of ancient amber was sliced and scanned to study the sand it captured when it flowed from a tree in what is now the Dominican Republic. **Fig. G:** Samples of light-colored sand were scanned against a black background.

Padre Island   White Sands   Florida Panhandle

For best results, scan plant samples as soon as possible after collecting them. If this isn't possible, preserve the sample in a cool location or immerse its stem in water. Leaves may be coated with wax, so be sure to clean the scanner's glass bed after scanning.

The color of scanned leaves may not appear true to life. If not, you can use a photo processing program to correct the color. I've found that the best way to accomplish this is to hold the actual sample next to the monitor and adjust the color of the image until it matches that of the sample.

Feathers

Most flight and tail feathers are easily scanned (Figure E), but white feathers and those with white down require a dark background. For example, I once scanned a flamingo feather. The upper half of the feather was pink, and it scanned fine. Because the lower portion of the feather was white, it was barely visible against the white background. The entire feather was visible when a sheet of black paper was placed over it.

Soil and Sand

Soil and sand samples are easily scanned. Light-colored samples require a black background. Scans like this are important for soil science, since they allow different specimens to be compared under identical lighting conditions.

As with all scans, the colors of scanned soil and sand samples might need to be adjusted. For example, when I scanned samples of light-colored sand from three locations (Figure G), the sand that looked whitest to the unaided eye was not as white when scanned.

Water can cause the color of soil and sand to change appreciably, and a scanner allows both wet and dry specimens to be scanned simultaneously.

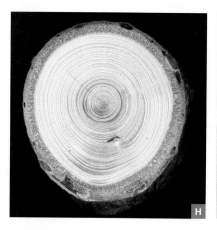

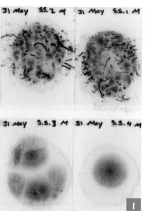

Fig. H: Growth rings in tree branches and small-diameter trunks are easily scanned. Shown here is a cross-section of a fir branch.
Fig. I: Colonies of bacteria and fungus spores can be easily scanned when they are grown on Petrifilm plates. These 2 films both show fungus colonies. The upper film was exposed to smoke from burning grass and the lower film was exposed to nearby clean air.

Tree Rings

In MAKE, Volume 19, I described how to scan the annual growth rings in tree trunks and branches so they can be analyzed with ImageJ photo analysis software. Briefly, use a sharp, fine-toothed wood saw to cut sections from branches and trunks. Sand the exposed face of a section with 100-grit paper followed by 220-grit. The final polish is made with 400- or 600-grit paper (Figure H).

Wood that was cut green should be allowed to dry for at least a day before sanding. You can enhance the appearance of the rings by moistening the face of the sample to be scanned.

Microbe Cultures

Bacteria and mold cultures are easily saved by digital scanning. My daughter Sarah's discovery of living microbes in biomass smoke arriving in Texas from the Yucatan Peninsula was made possible by exposing Petrifilms (made by 3M) to ambient air on days with and without smoke.

(See the article: Sarah A. Mims and Forrest M. Mims III, Fungal spores are transported long distances in smoke from biomass fires, Atmospheric Environment 38, 651–655, 2004, makezine.com/go/smokespores.)

To confirm her discovery, Sarah exposed Petrifilms to smoke from burning grass and to nearby clean air. The films exposed to clean air had only a few bacteria and mold colonies. Those exposed to smoke had dozens of colonies — plus black ash from the burnt grass. Scans of these films provided persuasive evidence of the presence of microbes in biomass smoke (Figure I).

Notes, Sketches, and Drawings

Field notes and sketches can be easily scanned for posting on the web or sending to friends, or to simply provide backup.

My ambition is to scan all of my field trip notebooks and photographic slides and prints and the thousands of log sheets on which I've recorded sun and sky data since May 1988. This will be a major chore that will require at least 10GB of storage. But the entire collection will fit on a tiny flash drive instead of a couple of heavy file cabinet drawers.

Going Further

Some specimens benefit from the inclusion of a scale in the scanned image. If the scale isn't perfectly straight, you can straighten the image in a photo processing program.

You can use free, powerful ImageJ software (download it at rsbweb.nih.gov/ij) to analyze your scanned images. (See my column in MAKE, Volume 18, for details about using ImageJ.)

Finally, there are many non-science applications for digital scanners that go well beyond making copies of documents and photos. For example, collectors of stamps and coins can easily digitize their collections.

Forrest M. Mims III (forrestmims.org), an amateur scientist and Rolex Award winner, was named by Discover magazine as one of the "50 Best Brains in Science."

Scan by Sarah A. Mims (Figure I)

THE MAKER'S BILL OF RIGHTS

If you can't open it, you don't own it.

Ease of repair shall be a design ideal, not an afterthought. **1**

2 Screws better than glues.

3 Consumables, like fuses and filters, shall be easy to access.

4 Torx is OK; tamperproof is rarely OK.

5 Batteries shall be replaceable.

6 Special tools are allowed only for darn good reasons.

7 Meaningful and specific parts lists shall be included.

8 Cases shall be easy to open.

9 Components, not entire subassemblies, shall be replaceable.

10 If it snaps shut, it shall snap open.

11 Standard connectors shall have pinouts defined.

12 Docs and drivers shall have permalinks and shall reside for all perpetuity at archive.org.

13 Circuit boards shall be commented.

14 Profiting by selling expensive special tools is wrong, and not making special tools available is even worse.

15 Power from USB is good; power from proprietary power adapters is bad.

16 Schematics shall be included.

17 Metric or standard, not both.

Illustration by James Provost

Maker

Transcendental Problem-Solving

MythBuster Adam Savage talks about making his way growing up.

Interview by Paul Spinrad

MAKE projects editor Paul Spinrad caught up with Adam Savage in San Francisco during production of *MythBusters'* ninth season. Earlier that day, Savage had found a local source for industrial quantities of citric acid, which he and co-host Jamie Hyneman needed to determine whether Alka-Seltzer and water could produce enough pressure to break out of a jail cell.

Paul Spinrad: How did you get started making things as a kid?

Adam Savage: As I said in my welcome to this issue [*see page 11*], my father was a painter, which gave me a real advantage. We'd see him paint a background, and it would have all this life to it, and he'd look at it and exhale sharply in a kind of silent whistle. As kids, it was a great thing for us to see him get excited about. When I wanted a race car for my teddy bear, he actually made one for me out of fiberglass. He made a chicken-wire frame and laid fiberglass over it.

PS: Wow — in the 70s that was cutting edge!
AS: Polyester, man, no epoxy back then. It smelled really bad. He sanded it, painted it, it was gorgeous. I loved that car, also knowing that he made it. My dad also helped me make a suit of armor out of tin when I was 13 for Halloween. We cut the pieces up with tinsnips and then fastened them together with rivets. Pop rivets — what a great technology! I wore the suit to high school and passed out from heat exhaustion.

When I wanted to make something, my dad said OK, here's the scissors and the

mat board and the masking tape. So I made a lot of things out of cardboard.

One day, I must have been maybe 13 or 14, I made a life-sized man out of cardboard. And that was the first transcendent moment for me, where I felt a release from all the cares of the world into the mental process of thinking through this project. At one point I stopped and went to my mom in the kitchen, and I told her something like, "I just want you to know that at 5:04 p.m. on this day, I am truly happy." I was being a bit dramatic, but I still remember that feeling so vividly. It was real joy. And the cardboard man sat out in front of my parent's house for a couple of years.

Another time, I found a refrigerator box after school and was pushing it home. This tough kid Peter came up to me and said, "This is my box." And I said, "No no no no no." So we had a push fight. I pushed him and he backed up and he said, "Didn't move me! Didn't move me!" I pushed him again, and I don't remember exactly how it played out, but he went away. I used that box to make a spaceship for an 8mm movie that my friends Paul Caro and Eric Pack were shooting. We even stop-motion animated an explosion. They tried to hold perfectly still in the shot while I pulled out like 2 feet of yarn to be the laser blast and then uncrumpled this explosion burst shape that I cut out of paper. I never saw the footage!

After that, I ended up moving the spaceship into our guest bedroom closet, which was like 4 feet wide by 12 feet deep. For the back wall of the closet I painted some cardboard black, punched some holes in it, and put a light underneath to make a star field. I put the cockpit about 4 feet back from there, so it was like you were looking out into space. It was so much fun making that environment and getting in there!

PS: So much learning comes from doing things that you enjoy, experiencing that feeling. And you've also talked about how failures stay with you, when you feel you've let people down. It's interesting how first-hand experiences guide you in a way that explanations and lectures never can.
AS: Right. When I was 19 or 20, I was bouncing around apartments in New York, living alone for the first time. I got this free apartment in Brooklyn, where the landlord was a collector of my dad's paintings. I set up a little studio space there for

making sculptures. I was lonely, but creatively it was a fertile couple of years for me.

I remember another transcendent moment from that time. I was making a piece that was like a phone from Hell. I glued all this stuff to a normal phone, making it like something from the movie *Brazil*. I decided to paint it in black and yellow stripes, but with all the tubes all over it, I had to spend an hour just masking. I worked on it until like 2 or 3 in the morning, totally getting into the zen of doing this thing, while this terrible Roddy McDowall movie was on TV.

Around that same time, while taking the subway home from my friend David's house, I started writing what I called "The Manifesto." It's 15 pages of furiously scrawled notes, and I remember looking up from my notebook once and seeing some guy on the subway looking at me like, what the heck is he writing? The Manifesto is basically my realization that being creative is like seeing your way above the clouds, to borrow terminology from Ram Dass. It's like seeking spiritual enlightenment. The real transcendent feelings you get are few and far between, but they keep you going to the next one. The cruel joke is, they get harder to get to, because you always have to go deeper. You know more about yourself, and you want more out of what you're doing, so you get better at it. The result is that everything you do well is the result of personal risk-taking and exploration.

PS: Doing something you already know how to do isn't going to get you there.
AS: Yes, and I don't think there's any separation between someone who's an excellent banker and someone who's a terrific painter. It all comes from the same emotions. I see creative endeavors as problem solving: you give yourself a problem, and while you're solving it you realize that you're miles from where you thought you were going to be. There are no concrete guidelines that can help you. You just learn how to dance with the thing that you're doing.

I think a lot of people get frustrated when what they start to make doesn't look like what they were thinking. The trick is, it's never, ever gonna look like what you pictured in your head. As Diane Arbus said, "I have never taken a picture I've intended; they're always better or worse."

"Everything you do well is the result of personal risk-taking and exploration."

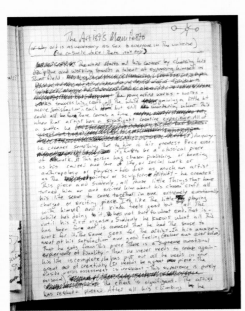

PORTRAITS OF THE MAKER AS A YOUNG MAN: (clockwise from top left) Savage, age 7 or 8, at a family friend's studio; at 19, directing friend David Bourla's film school thesis; at 19, carrying his plaster Thanksgiving turkey with alien popping out; the first page of his Manifesto, written at age 19.

PS: I think there are two frames. One is where there's a right answer, or you want to meet specifications, and you're successful if the result is how you imagined it ahead of time. And the other is, this is a process, I don't know where it's going, but I'm going to enjoy it and let it unfold.

AS: You can build things to meet expectation, like for a job, but the most creative work is when the material shows you something unexpected, or you get new ideas as you're working. There's a book that I love called *Interviews with Francis Bacon: The Brutality of Fact* by David Sylvester. It's a phenomenal book, and I don't know anyone else I've read in modern criticism that talks about things like Truth and Reality as terms of art that makes sense. Bacon describes painting as a process where you attempt to represent something, and this is true whether you're a representational painter or abstract or whatever.

And you always fail, but what you end up with is just as real as the original. A great portrait has a personality of its own, a vitality, and any artist will tell you that their best work changes every year. I look at favorite sculptures I've done years later and think, "How did I know how to do that?"

PS: One thing I'll notice while making things is long periods of time when no words go through my head. I just think on the level of what's in front of me, bypassing all symbols, which is grounding. It anchors you to reality. "Be here now."

AS: I see it as a form of meditation. Making things is a great form of escape, particularly in one's own shop. The radio's playing, you're making something that's good, and it works. It's incredibly relaxing and satisfying.

My favorite part is first getting my head around something, so that I understand it. I start out going, how the heck are we going to make a lead balloon? Then I think, we've got to spread the load, spread the force, do this, do this, all those things. Or when I'm making props that I like, the meditation I get into is, how do I make it really tight? How do I make this into something that I really want to have in my home?

I got into prop making after I moved to San Francisco in 1990. For my first three or four years here I focused on serious sculpture. I got some good press, had some shows, sold some pieces, and then I found my way into special effects,

which I had tried to break into before. But this time I was actually in the right place. I found I was good at it, and I wanted to put all of my creative faculties into it. So I gradually stopped making sculpture, at least in the fine art sense. I can only do one thing at a time.

I was working at Industrial Light and Magic when I started participating in the Replica Prop Forum [therpf.com], which I still visit almost every day. It's the central hub for a vast community of replica prop builders. There's an R2-D2 builders club, a C-3PO builders club, B-9 from *Lost in Space*, Robbie the Robot, the Dalek, plus costumes, scale models, paper props, and so on. I've posted lots of information to the RPF and I do lots of research there.

For example, there's been a great recent discussion thread about the Farnsworth Communicator, an alternative-history handheld videophone from the TV series *Warehouse 13*. Someone posted screenshots of it from their HD recording of the show. Someone else then takes measurements off the screen and says, OK, this looks like this part from Digi-Key. Then someone else says here's one that's slightly better, and everyone starts comparing them. Then people started creating and sharing graphics files. This thread is 55 pages long, with 115 comments to date, and pictures on almost every one. I'm not interested in making this prop myself, but I went through the discussion and was like, oh my god!

And then there are the prop collectors who pay lots of money for original props and they want the only one, or at least they want to control who makes accurate replicas. So some replica makers get information that they can't reveal. I myself have secretly distributed specifications to a couple of prop builders' groups based on information I had access to at the time. I won't say which forums, but I was supplying measurements, and the people who ran the forum just explained that the source was rock-solid, but they could not reveal who it was.

PS: There should be a declassify date for this kind of information!

AS: With some of the *Star Wars* props, people know so much that it gets to the point of absurdity, like, "The scope rings on that Han Solo blaster are incorrect; they should actually be ³/₃₂"

Savage's Dragon*Con Wookiee costume (above), built by Mark Poutenis. Savage dressed as a myth-busting Neo (right), jumping off a building.

"In these situations I remind myself what I'm doing is universal. I'm enjoying myself."

narrower!" And there are different philosophies. For example, the R2-D2 builders club now has sufficient detail to make a perfect R2-D2 — literally better than anything that ILM has: all aluminum, all remote control, everything. And there's a conflict in the club over how dirty and beat-up to make your R2-D2. The real thing is pretty crunchy. It's a piece of junk that's been reglued together many times on set. But some guys build these pristine R2-D2s, even though it's never been seen in the movies like that.

After ILM, I stopped working in special effects and started working for a toy company. The owner turned out to be Michael Joaquin Grey,

who's an amazing sculptor; he recently had a show at the Museum of Modern Art in New York. One time I told Michael that although I wasn't doing sculpture, I never wanted to say, "I used to do sculpture." And Michael said, "Come on, you're making your own R2-D2. What do you call that?"

PS: I was just going to ask that. I mean, art is called art because people agree to call it that. They agree that it has special value, and it's relevant to the culture. And that's exactly what happens in the replica prop community.

AS: Yes, but that's more blue-collar. It would have to be classified as outsider art. One of Michael's earliest pieces, which he got a lot of attention for, was a perfect 1-to-1 scale model of Sputnik. He called it *My Sputnik*, and I loved that so much that the first time I made a Maltese Falcon, from the classic movie, I called it My Maltese Falcon.

What separates Michael's Sputnik from my Maltese Falcon or R2-D2 is that he brings a different context to it. As an established sculptor, he opens it up to a broader cultural discussion with the world. It might be said that I'm creating the same kinds of objects, but I'm thinking how cool it would be to have my own droid.

I'm finding that the places where these communities come together and define themselves are the cons: Worldcon, Comic-Con, Dragon*Con. That's where many people who collaborate remotely finally meet in person. I've got friends from the RPF I've worked with for years but have never seen a picture of, and they'll fly out to one of the cons and spend a weekend with me. I'm so enthusiastic about the whole experience. I'm going to Dragon*Con over Labor Day, and I've got my costume all set up. I'm looking forward to it!

Wearing costumes is another transcendent experience for me: putting together something that's lovely, a character that I like. I'm also a little embarrassed about it, but therein lies the thing, you know?

PS: Where does the embarrassment come from?
AS: Well, adults dressing up is weird. There are times on *MythBusters* where I surprise people by walking out in a costume that I've been planning for a while, and then I'm embarrassed at how much I'm enjoying it and how juvenile it is.

We did a show recently where I jump off a building. I don't want to tell you the myth, but during my training for the jump I wore a sweat-suit that said "Trainee." At graduation, I thought that if I'm going to be jumping off a building on high-speed camera, I should be dressed like Neo from *The Matrix*. So I got some knee-high, buckle-up boots from a fetish store on Haight Street, bought a long coat on eBay, got some jeans with a climbing harness on like he had.

And when I walked out, I could tell from the crew there was a little bit of, "Oh, here's Adam wearing one of his costumes again." And I felt a little embarrassed, but I also knew that jumping

off a building in this thing was gonna be awesome. As soon as we saw the high-speed shots, with my coat flowing back behind me, my director turned to me and said, "You were so right about that outfit. This is a monster of a high-speed shot." Here, I have a shot, I'll show you.

PS: Oh yeah, that's perfect, totally Neo.
AS: But I'm wearing my own hat. In these situations I remind myself what I'm doing is universal. I'm enjoying myself. I'm pretty honest with the camera, and the person I am comes across, which is a skill in itself. With that skill, I want to inspire kids to be confident that they can do what they want and not worry about embarrassment. I think that's a lovely message.

That's one of the reasons that I tweet so much about going to the cons and putting on costumes, about making stuff and being proud and excited about the stuff that I make. I think it reflects a universal truth that people will identify with. It's one of the things that I've got to give.

PS: I think of enthusiasm as the opposite of coolness, and adolescence is a turning point for this. Children are all enthusiastic, they love what they're into and that's it. But then something happens, and suddenly some of the kids start looking down on that enthusiasm and seeing it as immature or dorky. So they invent coolness as an alternative. I always gravitated away from that because I was interested in too many things.
AS: Yes, and enthusiasm also makes you vulnerable. When you like something, someone can take it away from you. I once gave a sculpture to some friends as a wedding present, and they turned it down. That was really upsetting to me. And that vulnerability itself is also embarrassing. The two emotions are deeply linked, which is why people try not to cry in public.

PS: Growing up, what were some of your main cultural influences?
AS: A lot of *Mad* magazine, movies like *Star Wars* and *Raiders of the Lost Ark*, *Fangoria* and other movie magazines. *Fangoria* always had something about homemade projects, homemade horror films. I remember I took a lot of magic books out of the library.
PS: Yes, me too. Why do you think magic becomes

Savage: "I'm an instrument fetishist. I love measuring (and being able to measure) *everything*. This is nearly every instrument of measurement I own in one case. Since these photos were taken, I've added five more."

so important at that age? What's going on, in terms of learning the shape of the world?

AS: In *The Liars' Club*, Mary Karr writes that the first time you tell a lie and get away with it is the first time your adulthood steps through you. That is when you realize that your parents are not invincible. And there's nothing more powerful or lonelier.

Magic is also a way of learning how to capture someone else's attention. I've been playing with it more lately. Performing magic starts with the trick itself, but it's more about maintaining a tension in the audience between the possible and the impossible. A good magician would never tell you that something they do is truly impossible.

PS: That's it. I remember the same time I got into magic, I loved *Ripley's Believe It or Not!* and *The Guinness Book of World Records*. Looking back, I don't think I was necessarily amazed by all the records, but I did want to learn where the edge was between possible and impossible. It was very educational to learn that, OK, people can't be more than about 1,200 pounds. That helps you build your model of the world.

AS: But the *Guinness Book* we had when we were kids is very different from the one my kids have access to. Now it's a big, glossy volume, not the little paperback with impossibly small type and dark photos of things like Largest Goiter.

PS: As a dad, what's your take on education these days? I think our system pressures kids toward learning things that are abstract and

symbolic, and of course more standardizable.

AS: You mean as opposed to the things like drama club and the shop classes? Yes, and when education budgets are cut, those are the things that fall by the wayside — all the things that are the most fun. For me, the shop classes, art classes, and especially the drama club were absolutely critical.

This past summer I sent the kids to different camps, including acting camp and rock 'n' roll school. They got exposed to such interesting people doing interesting things, and I wish they could get that in school at least once a week.

The main thing is, you want teachers who are interested in their subjects. Those were my best teachers. I recently took my kids to a skate shop because they needed some new parts for their skateboards, and they got into this deep conversation with the guy behind the counter, in this whole language about skate parks and people who skate, and all this stuff that I don't know. It was lovely watching them absorb it all.

There are always arguments about what school should be for, if it isn't just state-sponsored babysitting, and we always wind up thinking of the most utilitarian thing, what's the bare minimum we can do. But in utilitarian terms, the only really viable thing I got out of 12 years of education was typing. Kids are told that they have to sing on key, but what we should really be showing them is how to enjoy singing, and the on-key stuff will happen later.

Paul Spinrad is projects editor for MAKE magazine.

Photography by Cody Pickens

Toy Story

How the creator of the Superplexus turned a childhood idea into a lifelong passion. By Michael McGinnis

While visiting my family in Colorado in 2002, I came across a captivating object sitting on a shelf in my sister's house. It was a cantaloupe-sized plastic sphere that housed a labyrinth of purple and turquoise colored ramps, tubes, and drops. It had a small steel ball rolling around in it. I needed no instructions to tell me that this was a puzzle, the object being to roll the ball through the three-dimensional maze from the starting point to the finish without having the ball fall off the track.

I thought it would be easy to solve. A week later, I was still spending a couple of hours a day trying to guide the ball to the end. But instead of feeling discouraged and frustrated, I felt challenged and encouraged to keep trying.

The thing was called the Superplexus and was

made by Tiger Electronics, a division of Hasbro. I didn't solve it until weeks later, and when I did, I re-challenged myself by seeing how quickly I could complete the maze. After a few months, my sister and her kids wanted it back, so I had to part with it, but I never completely forgot about it, and in 2007 I went online to order one for myself. I was pleasantly surprised to discover that the creator of the Superplexus, Michael McGinnis, had a website, and that he lived in Santa Rosa, Calif. (a 15-minute drive from MAKE's office) where he taught 3D design at a college. I contacted Michael and we started a correspondence. I asked him to write about the origins of the Superplexus for this issue of MAKE. This is his story.

—Mark Frauenfelder, editor-in-chief of MAKE

Photography by Rebecca McGinnis and Michael McGinnis (far right)

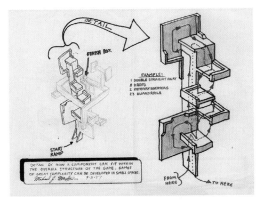

Having an idea is easy. Drawing it out can be challenging. Making it physically is an altogether different experience.

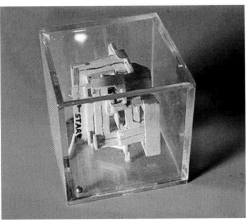

ROUND IT UP: Superplexus creator Michael McGinnis (far left) pictured with his Giant Superplexus model; (this page, counterclockwise from top) an early sketch and the finished model of the Equilibre Hable, precursor to the Superplexus; the new Perplexus toy.

uperplexus evolved from a project given to me 31 years ago by my 11th-grade art instructor, Ed Hairston. His project: design a board game.

I couldn't follow directions very well and had trouble reading and studying, or even knowing what was really expected of me. For example, in chemistry, we were asked to calculate the thickness of aluminum foil. While the other students used atomic weight ratios to measure the foil, I simply folded the piece over and over again until I could measure it with a wooden ruler, then divided that thickness by the number of folds. (Of course, I had to crush it in a vise to remove the empty space to get an accurate result.) I now realize that a hidden learning disability provided me with a unique way of looking at things.

Art Class Assignment

Since I was bad at board games, and felt especially terrible about losing or even beating others at them, I decided I'd rather design a maze. My family had just relocated to California from Illinois, where my friend Dale Lomelino and I spent innumerable hours creating complex mazes. In coming up with an idea for Mr. Hairston's assignment, I initially drew a design for a Marble Race Game. It was to have four

colored marbles in four side-by-side tracks. Players would compete to see which marble could make it to the bottom first. How would one release all of the balls at once? How would the winner be determined? How can a curved track guarantee that all four tracks are the same length? What a boring idea!

So, I flipped the paper over. I had a thought: "How can I make a 3D labyrinth with a BB running around in it?" I began to sketch and realized that it couldn't be like the popular maze game Labyrinth, with holes as obstacles, or it would be impossible to see. I drew a cube, and along the inside surface was a continuous road — an aqueduct/bridge/Great Wall of China-like path. A cup was at the end for the ball to rest in. There was no top or bottom; it was relative to how you turned the cube. I presented the drawing to Mr. Hairston, who directed me to make a model.

The model was constructed of balsa wood and glued with Elmer's Glue-All. I called it Equilibre Hable (meaning balance skill). It was a hodgepodge of ramps and railings in a clear, 3-inch plastic cube and was both hopelessly ugly and instantly intriguing. It was so difficult that I still have never gotten to the end! The school principal borrowed it for a month or so, and this model became my family's pastime at nearly every gathering for years to come. (Twenty-three years later my student Matt Wong became the only person to finish Equilibre Hable, and he did it on his first try.)

No Traction

I was convinced that this thing could be a real product. All I had to do was get someone to grasp the concept and run with it. Someone in the toy world would take it on, design a beautiful and elegant version, manufacture and market it, and I'd just watch it all happen. How wrong I was!

Two years passed before I dared to research toy manufacturers and make calls. No one even took the time to look at it. In hindsight, this was a good thing. I didn't have the skills or maturity to take the game to the next level. I was fearful of making a new design and physical model. At this point I could only draw up ideas. I did discover one thing: the game was about a concept, not a particular design. It was ramps, railings, and turnarounds. This idea was far too open-ended for me to solve at the time.

I can only describe the next phase as the lost years. I was in college working toward a degree in sculpture when a feeling came over me. How could I contribute to the mass consumerism and greed of our culture by making a commercial object? Think of all the resources used and the pollution that would result from being part of the Big Problem. Eventually I came to my senses, realizing that this was a thing of joy, and, if made properly, would not end up crowding landfills or washing up on beaches.

Refinements and Disappointments

Nine years had passed since I'd made the first model. Equipped with a master's degree in sculpture, I began teaching at Santa Rosa Junior College. I had time to work on the idea again. I did lots of theorizing about the essence of the game, and I had become much better at drawing and aesthetics. My new bride, Becky, encouraged me to pursue Psychopath (its new name).

I finally began to better define the basic elements: on-ramps, straightaways, drops, pathway inverters, complex inverters, single- and double-sided ramps, angles, guard rails, tunnels and tubes, transfer rails, and finish boxes. Injection-molded plastic parts could be cemented together.

Within a year, we fell victim to one of those invention submission scams and spent good money hiring a company to generate a marketing/manufacturing report. Armed with this useless information, I got

nowhere with the game. We hired a consultant who gave us marketing ideas that went nowhere. Again, it was a good thing, because my design sense was still too unrefined.

Another ten years passed, during which we had two children. It was 1997. I continued teaching art courses, and pursued patents and business prospects for mat cutter designs and picture framing systems as a way to earn extra income. This five-year exercise ended up being successful (only in my mind) because I learned about confidence and how to work with industry. Freshly freed from working on those inventions, I was eager to delve once again, deeper, into the game.

Breaking into the Toy Biz

Out of the blue, I asked my digital arts student Erin Montague if she knew anyone in the toy industry. "My brother!" she exclaimed. Erin's brother introduced me to Dan Klitsner and his team at KID Group in San Francisco. KID was well known and respected in the toy industry as inventors. They license ideas to all the major players.

Over the next two years, I worked on a series of cube models, refining concepts, while KID pursued contacts and offered advice on design issues. Dan would say, "We have a meeting with X tomorrow. Can you make a new model by then?" I'd stay up all night, travel to San Francisco in the morning, and drive back to my class in the afternoon. This was an awesome time.

KID made a deal with Stewart Sims (the guy who brought Rubik's Cube to the free world) to make and market Plexus, the latest name for my game, which then became Perplexus. Sims was with a startup company called Next Electronix. KID gave me two days to devise a new 4-inch spherical model, and three days to construct it.

Next Electronix was suitably impressed, and we collaborated for the better part of a year on tooling, prototypes, samples, production models, and packaging designs. We produced a TV commercial reminiscent of *That '70s Show*, with kids playing the game in their basement. I flew to New York for the American International Toy Fair, where the game was officially released. How exciting!

Superplexus Goes Global

Well, the toy fair was a great success, but Next Electronix was not. They went belly-up because of internal problems. The game ended up in boxes somewhere in a subsidiary of Playmates Toys. It

RAMP IT UP: Before any drawings, a full-scale foamcore model (above) was made to work out the design for Superplexus Vortex. Next came the CAD drawing for use with the ShopBot. Below, McGinnis and James Yonts worked jigs to hold the structure in different configurations; McGinnis carved a rubber arrow stamp, applying almost 1,000 arrows to guide the way, then applied the rails that would hold the ball.

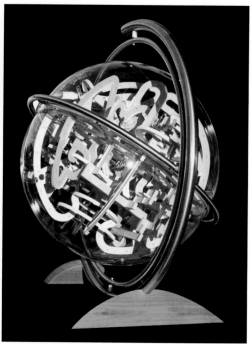

ROLL IT OUT: The internal structure, nearly complete (left). The final model encased in plexiglass (above). McGinnis' sister Mary (opposite) playing with *Superplexus Vortex* at the Sonoma Valley Museum.

took a while to get the rights back. Freshly wounded (but secretly relieved because I was not happy with the results), I built a very involved 8-inch spherical version unlike anything that came before. It was the first version of Superplexus.

KID loved it, and asked that I make a new design that could be manufacturable as soon as possible because they were meeting with Tiger Electronics (Hasbro) in a week and wanted them to see it. One hundred hours of work later, I had a design. It was too difficult to draw, and could only be designed by making it. Tiger went for it! I worked with engineers and CAD designers almost nonstop. It was the most intense, work-filled, rewarding six months of my life.

By the spring of 2002, Superplexus hit stores worldwide. It sold more than 700,000 units in a four-year period. It was listed in the "Top Ten Christmas Gifts of 2002" in the United Kingdom. It did quite well in Japan as well, where it was packaged with stickers in the shape of eyes, to look like a brain.

Killed and Revived
Despite the game's positive reception, the quantities sold were less than Hasbro's vision of success, and Superplexus was soon after discontinued.

Recently, we discovered several Chinese counterfeit versions; one even made it to the White House, given as a gift to Malia Obama during a visit to Pixar. One manufacturer, Buksi, has agreed to pay royalties to KID and myself in order to market a version in the United States as Perplexus (perplexus. net), distributed by PlaSmart. So you'll soon see Superplexus available again (sans electronics).

Giant Superplexuses
The latest phase in my creative endeavor is the Giant Superplexus. I built my first 4-foot diameter version for the 2007 Maker Faire in San Mateo, and it was by far the most complex version to date. This design had little in common with the 7-inch production model; it's got so many circular shapes in the design that I've dubbed it *Superplexus Circles*. Making the interconnected pathway system took approximately 100 hours, a difficult and challenging experience.

This model has since been exhibited in many venues, with amazing response. Thousands of people have enjoyed seeing it, although only a small percentage get a chance to actually play. I've tried to limit playtime to around 5 minutes unless the

Photography by Laura McGinnis (left) and Michael McGinnis

Photograph by Frank Hummel

player does not fall off the pathway. (If they're that good, how can I tell them to stop?) Two people have actually gone from beginning to end without falling off, taking more than an hour to do so.

I recently completed my first Giant Superplexus commission, for the Sonoma Valley Museum, called *Superplexus Vortex*. I used a completely new pathway design. It's enclosed in a 36-inch acrylic sphere, and is made from a variety of materials, including high-grade plywood, hardwoods, metals, and plastics. In addition, a major museum is contemplating a commission, along with several other venues.

Giant Superplexuses are complex works of interactive art, and I am very fortunate to have them as my most personal creative outlet. This art form stretches my mind while providing joy to others. Although I am adept at other media and concepts, no other pursuit holds such an iconic place in my heart. Here I feel as though I am truly contributing to the language of the world.

Word to the Wise

What I've learned from my experiences can be boiled down to a few typical life lessons: nothing worthwhile is easy, or ever turns out as expected, or comes out

of thin air. Having an idea is not the same as making the physical thing. The best way to solve a problem is to build it. Setbacks are just another step forward, perhaps along another path. Not all ideas are good ones. Ask and accept help from others. There is always room for improvement. There is no such thing as the perfect solution. Although money can't solve all problems, it sure does help.

And, when you believe enough in something and are willing to work toward a goal, even the seemingly impossible can be accomplished.

Do you want a Giant Superplexus of your own, or for a museum or other public venue? Do you wish to collect limited-edition, handheld Superplexuses? Would you like to exhibit them in a gallery? If so, let's talk! mmcginnis@santarosa.edu

---

Michael McGinnis is an artist/maker who enjoys making interactive works: furniture, digital art installations, Superplexuses, inventions, and more. Michael teaches sculpture and design, and is the Art Gallery Exhibits Specialist at Santa Rosa Junior College.

Maker

Mayor of Britannica

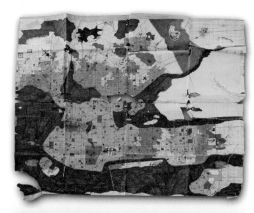

For 40 years, Michael Chesko has been mapping the country of his dreams. By Robyn Miller

Michael Chesko is building a country. As it turns out, this sort of thing takes roughly a lifetime. He began conceptualizing and mapping it in junior high school, and today at age 51 the Mesa, Ariz., father of five boys says his country, as yet unnamed, is "still in progress, it is indeed something I may never finish." How could it be otherwise when, after all, his nation is roughly half the size of Australia?

He calls this country his magnum opus. Every border, boulevard, city, street, and structure: it's all there. Rolls and rolls of maps, models, and flags, all of which have been shoved away into closets.

Astonishingly, the only people to ever get a peep of his dusty republic have been Chesko's friends and family. He is honestly bewildered and, I think, thrilled when I express interest in his masterwork. As if I've pulled a plug on something bottled up since 1970, he sends me map after map — a mere sampling of the hundreds that constitute his world, each one minutely rendered and congested with near-microscopic detail.

I would be remiss if I didn't mention Chesko's unparalleled megalopolis: Britannica. With boyish pride, he boasts of its vastness: "I have drawn more streets and freeways for Britannica than the whole Los Angeles metropolitan area has in reality." He knows it's true because he diligently copied L.A. maps to the same scale and spliced them together, "and Britannica is considerably larger." This must have been no small task; the Britannica maps, when laid across Chesko's carpet, gobble up an obscene portion of his living room.

The obvious question, about all these years of masterminding the layout of a nation, is why? Is it hobby? Art? Obsession? Why spend 40 years working on a world that may never be completed?

Chesko says that finishing is beside the point:

NATION BUILDER: Chesko drew this early map of Britannica at ages 13–14. Lately, he's been sculpting the city's downtown, aka Proto Britannica. He's also sewn 12 flags for his imaginary country's districts.

the fun is to imagine, to build, to perfect, and finally, to visualize himself within his creation.

"Britannica is a playground for my avatar, my fantasy self," he explains. "I go there and I live there. I walk around its streets and I experience it in every way I care to imagine. … It is the pursuit to perfect this experience that drives me."

➕ To see Michael Chesko's incredible carvings of Manhattan, visit makezine.com/go/chesko.

Robyn Miller is obsessed with miniatures. He also enjoys occasionally painting, writing music, and directing video games. tinselman@gmail.com

Iron Man

Japanese blacksmith Kogoro Kurata forges ahead.

By Lisa Katayama

POP IRON: About 20,000 fans mobbed the unveiling of Kurata's giant anime robot. Others enjoy his wrought-iron decor at a hip Tokyo pizzeria.

What do a modern pizzeria, a vintage keyboard, and a giant anime robot have in common? For most of us, nothing at all. But for Kogoro Kurata, they're pieces of a growing portfolio of unusual things he makes with iron.

A second-generation blacksmith, Kurata has been hammering, bending, and forging iron since childhood. While he's done his share of building fences and gates to make ends meet, he is best known for his intricate, off-the-wall sculptural creations.

"Iron is like Play-Doh at high heat," he says. "You don't need to plan ahead — you can make things up as you go along. Plus it's much safer than playing with wood!"

The dark, rustic look of Kurata's ironwork is a far cry from the colorful kitsch that decorates much of his hometown of Tokyo, but that hasn't stopped Kurata from gaining popularity. At 17, he used iron from his dad's workshop and made a playable bass guitar. That went well, so he made a violin and a cello, too. The trio of instruments won him an esteemed local design award and a steady following of fans.

Since then, Kurata, now 36, has made dozens of random items using metals, including opera sets, giant flowers, and a wrought-iron workstation for his Mac. His most famous work to date is a life-sized re-creation of the robot Scopedog from the cult classic anime *Armored Trooper Votoms*.

"I wanted to do something completely ridiculous," he says. "Scopedog is an anime robot, but it looks like something you could buy from a street vendor." The sculpture may have been just a fun project for Kurata, but it was cause to celebrate for *Votoms* fans — 20,000 showed up at its unveiling in 2005.

Commercial enterprises have tapped into Kurata's talent, too. Seirinkan, a pizzeria in a high-end Tokyo neighborhood, is one of Kurata's newer designs — opened in 2007, it has an intricate decor with hundreds of tiny, handmade, hinged square windows and a wrought-iron spiral staircase that snakes up its four-story spine.

The building's camouflage-net-covered façade is an anomaly on the otherwise nondescript one-way street. Kurata was initially hired just to make its gates, but then the owner asked him to revamp the whole place. "I don't want it to look like an ordinary pizzeria," the owner said to him. "I want you to let all that Kurata-esque gooey hellishness come out."

Kurata carefully documents all his work on a blog called *Nandemotsukuruyo*, which is Japanese for "I'll make anything." Each post tells the story of how he made something and why — under "I wanted to live in it so I made it," he writes about the dome-shaped soccer ball house he designed and built from scratch with $10,000 at the age of 22.

"For me, making is about things you want to try, things you want, things others don't have," says Kurata, who's lived in the soccer ball for over a decade. "The most important thing is that I'm having fun."

» Kogoro Kurata's ironworks: ironwork.jp

Lisa Katayama writes for *Wired, Popular Science, New York Times Magazine*, and her own blog, TokyoMango. She's a contributing editor at Boing Boing and the author of *Urawaza*.

Be an Angel

Take a school or teacher under your wing this holiday.

Right about the time you're reading this, a bevy of makers in Sebastopol, Calif., are being sustained by pizza and beer in a warehouse as they burn the midnight oil fulfilling the upswing of makershed.com orders we always see this time of year.

Even last year, when most retailers were feeling the cold sting of recession, the Maker Shed's orders were up 72% over the prior year. I mention this not as a boast, but rather because we see it as an indicator of a bigger trend taking shape. A making trend.

Not only are people giving gifts they've made themselves in greater numbers than ever before, but they're also touching loved ones with the gift of making, by giving a project kit that enables the recipient to make something of their own. Passing the DIY torch, if you will.

This holiday season, however, I want to challenge our readers to extend the gift of making beyond your friends and family. I want you to consider giving a gift to a school, a science class, a teacher, or maybe an after-school mentor at a school near you. Spread your wings. Become an angel to a teacher or an after-school program.

Regardless of where you live, I'll bet there's a school within walking distance that would love to be able to stimulate the minds of students with a good hands-on science project — a DIY electronics kit, a basic laboratory equipment kit, a robot kit, or a Blinkybug. Even a gift of $20 can make for a fun class project.

Here's my pledge to you. When you place an order with makershed.com and name a school or teacher as the recipient at a bona-fide school address, we'll take 10% off the entire order. If your order is $250 or greater, we'll take 15% off the entire order and include *The Illustrated Guide to Home Chemistry Experiments* as an additional gift. Just enter code "School10" when placing any order under $250 and specify a school and/or teacher at a school address. Enter "School15" if your order is $250 or greater.

Playing with the Snap Circuits Flying Saucer Kit.

Whether you buy from Maker Shed or some other fine supplier, or build it yourself, whether you're thinking about your own child's class or don't even have kids, please consider spreading the maker ethos and spirit to a school or teacher near you.

Special Announcement

We've launched the **Make: Science Room**, a hands-on science destination for backyard scientists and educators with how-tos on setting up a home lab, evaluating and buying equipment and supplies, and doing fun and educational home science experiments.

We've also put together amazing bundles of lab equipment, tools, and chemical sets at some outstanding prices. The Maker Shed now carries everything from high-quality microscopes, to all manner of fancy beakers and flasks, to lab aprons and splash goggles. Check out all the how-to science goodness at blog.makezine.com/science_room.

Happy Holidays,
Dan

Dan Woods is associate publisher of MAKE and general manager of the Maker Shed.

Photography by Shawn Connally

SCIENCE ROOM

All LAB, NO LECTURE

Make: SCIENCE ROOM

Our growing **Science Room** carries the perfect gifts for the home scientist. From microscopes to chemistry sets, we've got you covered, aprons to zinc.

MICROSCOPES
Seven high-quality models, starting at $120, suitable for elementary through graduate school, with a wide range of supplies and accessories.

BASIC LAB EQUIPMENT KIT
Specialty equipment you need to set up your first home lab, from beakers to burners, test tubes to splash goggles.

MAD SCIENCE
55 experiments you can do at home (but probably shouldn't) from a homemade hot tub to cooking with liquid nitrogen.

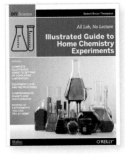

HOME CHEMISTRY
How to set up and use a home chemistry lab, with step-by-step instructions for conducting experiments.

THAMES & KOSMOS CHEMISTRY SET
From a mini fire extinguisher to invisible ink, more than 350 fascinating and fun experiments are included.

THE ELEMENTS OF SCIENCE KIT
100 experiments in biology, chemistry, and physics. Includes electromagnetism, the human body, and optics.

Maker SHED
DIY KITS + TOOLS + BOOKS + FUN

MOUSEBOTS

MECHAMO KITS

STEGOMECH

LED LIGHT BRICK

MINI MONSTERS

BLINKYBUG KITS

Welcome to the Maker Shed. In the following special holiday section, you'll discover just a few of the unusual and hard-to-find DIY projects available for purchase in our makershed.com store. Carefully screened by the editors and staff of MAKE magazine, these Maker Shed kits, tools, books, and games are designed and produced by our favorite makers and small suppliers from around the world. Our goal is to offer life-enriching challenges and exploration through carefully curated science, tech, and crafting projects for a wide range of interests and experience levels. We hope you enjoy them!

HAPPY HOLIDAYS + PERMISSION TO PLAY!

INTRO TO ELECTRONICS

Resistance is futile! Unless you know which resistors to use. From basic soldering to designing electronic circuits, we'll help you get started on the road to technological domination.

MAKE: ELECTRONICS BOOK & KITS
Our brand-new, hands-on electronics book makes learning easy, entertaining, and fun. Pick up basic electronics tips and techniques with step-by-step instructions, circuit diagrams, full-color photos, and engaging explanations of the principles behind each project. Check out the MAKER SHED COMPANION KITS designed to help you with the projects in the book!

DIY ELECTRONICS KITS
Harness the power of the electron with this component-loaded kit. Everything you need to get started is in the box.

LEARN TO SOLDER KIT
Includes soldering iron, instructions, and everything you need to solder together a simple but satisfying project.

MAKE BEETLEBOT BUNDLE
A great starter robot. Includes the robot guts and a copy of MAKE, Volume 12 for instructions.

THE BEST OF MAKE
A surefire collection of fun and challenging DIY projects.

HAYWIRED
From a talking greeting card to a Mona Lisa that smiles wider when you approach, follow step-by-steps to build unusual contraptions.

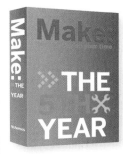

MAKE: 5TH YEAR BOX SET
The four volumes of MAKE's fifth year are combined in a Special Edition Boxed Collector's Set. Includes Volumes 17, 18, 19, and 20.

« AND DON'T FORGET BACK ISSUES!

STOCKING STUFFERS

3D CHRISTMAS TREE
The ultimate Christmas gadget features 16 flashing LEDs, and comes together in 8 simple steps.

LED MENORAH
Perfect for Hanukkah, this mini-menorah displays one more light each time it's switched off and on.

DIY LED ORNAMENTS
Light up your tree with these LED felt ornaments. Choose from a dove, reindeer, or snowboarder. No soldering necessary!

MINTY BOOST
Small and simple but powerful USB charger for your iPod, camera, cell-phone, or anything you can plug into a USB port to charge.

SOLAR-POWERED THEREMIN
Flying saucer music, powered by the sun! Easy-to-build kit for electronic music.

POCKET REFS
Choose from a variety of pint-sized reference books, chock-full of need-to-know info. You'll never be stumped again.

LEATHERMAN TOOLS
Small enough to fit on your keychain, these are the perfect companions for mobile fixing, hacking, and MacGyvering.

SOCK MONKEY
This creative starter kit will have you making an adorable sock monkey faster than you can say "eek eek."

TV-B-GONE
Turn any TV on or off with the click of a button. Peace on Earth is at your fingertips.

DISCOVER DIY KITS, TOOLS, BOOKS, AND MORE AT makershed.com.

Make: Kids

Inspire a kid (or your own inner child) with these out-of-this-world projects.

We've looked far and wide for the very best kid-friendly projects in the makerverse. Behold rocket mega-launchers, secret treasure boxes, magical marble adders, and the wooden sailboat you've wanted to build since you were 10. Revel in the golden age of maker toys, and discover the projects teachers love best.

Illustration by Adam Koford; photograph by Cody Pickens

Make: Kids

<table>
<tr><td>

Productive Plastic Playthings

</td><td>

A LOOK BACK
AT 1960S
MAKER TOYS.

BY BOB KNETZGER

</td></tr>
</table>

The 1960s were a golden age for toys in America because of a timely combination of postwar factors. The baby boomers were in their prime toy-buying years ("for ages 6–12") while the boom in the economy created more purchasing power for their parents. For the first time, toys were sold using network television ads, launching nationwide fads. The graduate of the 60s was told there was a "great future in plastics," while at the same time the Space Race spurred an interest in science education and technological toys.

Among time-tested playthings, like dolls and trucks, came a new category of toys: "make and play." There had been creative kid crafts before, like paint-by-number kits or Erector sets, but these modern maker toys inspired a generation of kids to mass-produce their own creations using miniature, at-home versions of industrial manufacturing methods and advanced "space-age" materials.

With clever research and development and bold marketing approaches, Mattel Toys led the way in make and play. Their innovative products included the Vac-U-Form, Thingmaker, and many other cool maker toys.

Designed in 1963 by whiz inventor Jack Ryan, a Yale-educated guided missile engineer (and Zsa Zsa Gabor's sixth husband — how's that for an only-in-the-Swingin' Sixties resume?), Mattel's

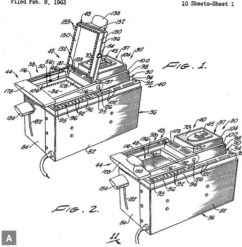

April 27, 1965   J. W. RYAN ETAL   3,179,980
TOY FORMING APPARATUS

Filed Feb. 8, 1963   10 Sheets-Sheet 1

FIG. 1.

FIG. 2.

A

MOLDS PLASTIC...MAKES MOST ANYTHING!

VAC·U·FORM

IT'S FUN! IT'S USEFUL! NOW YOU CAN DO MOST ANYTHING WITH PLASTIC! HERE'S A WONDERFUL NEW WAY TO MAKE YOUR OWN TOYS, GAMES, GIFTS, AND MANY OTHER THINGS!

BY MATTEL

B

C

VACUUM PUMPED: Figs. A and B: Inventor Jack Ryan's Vac-U-Form allowed kids to mold plastic into 3D toys. Fig. C: Each kit supplied more molds, more projects, and more sheets of plastic, including both transparent and metallized chrome in gleaming colors. Fig. D: After 3 minutes on the 110-volt plug-in heater, the softened plastic was stretched and formed over a mold with a few quick strokes of the pump handle.

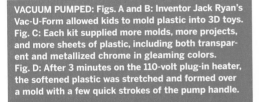

D

Vac-U-Form used air pressure to mold small squares of colorful polystyrene into three-dimensional shapes. Vacuum-forming thermoplastics, the process used by Boeing to make jet airline interiors, was now at the fingertips of kids across America.

And those fingertips often got blisters! Unlike today's super safe, car-seated and bike-helmeted kids, baby boomers braved the dangers of toys like the Vac-U-Form. Its exposed heating plate reached the skin-sizzling temperature of 350°F. (You can tell it's Mattel, it's … *Owwwww!*) No matter. With the awesome power of the Vac-U-Form, kids cranked out cars, planes, signs, disguises — all kinds of mini toys using the many molds that were available.

Just as innovative as the Vac-U-Form's design was its marketing. Like Barbie with her never-ending array of clothes (sold separately!), the Vac-U-Form line of toys had dozens of accessories and refill kits

for making jewelry, medals, badges, airplanes, animals, boats, military vehicles, and more. A nation of kids was mesmerized by the Vac-U-Form's TV commercial featuring the magical moment of transformation: molded shapes took form before your eyes, set to the ba-boom beat of timpani.

A familiar smell can trigger a flood of childhood memories, and just one pungent whiff of burning plastic is enough to evoke the Thingmaker, a spinoff of the Vac-U-Form. The same heater that softened stiff sheets of styrene could be used to cure liquid plastisol. This goopy mix of polyvinyl chloride in a solution of plasticizers is used to manufacture soft parts like tool grips, squeezable coin purses, and flexible fishing lures. Renamed Plastigoop and packaged in handy squeeze bottles, the protean plastic came in a dozen colors (including the exotic glow-in-the-dark formula). The Thingmaker's molds were

Make: Kids

PAKS FOR MAKERS: Fig. E: Fighting Men molds for the Mattel Thingmaker. Watch out — they get hot! Fig. F: Maker-Paks for Thingmaker and Vac-U-Form machines. Fig. G: No one who used Plastigoop can ever forget its distinctive aroma. Fig. H: Pre-computer pixel art.

made from die-cast zinc, the same metal used in toy cap guns, which reproduced each tiny mold detail, from the hairs on a spider to the gruesome stitches on a shrunken head.

Like its older brother the Vac-U-Form, the Thingmaker offered the excitement and danger of high-temperature fabrication techniques. The fun began by filling the mold with various colors of Plastigoop. Heating the mold on the Thingmaker's oven cured the Plastigoop into a wiggly gel. The last step was to quench the finished mold in a pan of water, making a satisfying blast of steam. The only concession to safety was the wobbly wire handle used to lift the smoking-hot molds.

There were many different themes for Thingmaker's "Maker-Pak" sets, but the most successful was Creepy Crawlers. The TV commercial (it's on YouTube) featured a James Mason sound-alike voiceover drolly describing how to scare Mom and annoy newspaper-reading Dad with the wiggly worms and rubbery bugs. Mattel went on to produce Giant Creepy Crawlers (with flocking to make the bugs fuzzy), Creeple People (cute-ugly trolls with interchangeable heads, arms, and legs

that connected together on a pencil), Fighting Men (army men with wires for bendable arms), and Fright Factory (shrunken heads and skeletons).

Other toy companies sold similar items. Kenner's 1964 Mold Master boasted that it "makes solid — not vacuum-formed toys." Boys could mold army men and mini toy guns that shoot, while girls created small dolls and accessories, including multicolored "miniature teenagers at a birthday party" with tiny bongo drums, record players, and soda glasses.

Perhaps the weirdest heat-powered plaything was the Strange Change Time Machine. Instead of merely molding raw plastic into mini toys, this clever contraption provided endless play with 16 different time capsule creatures. "Create 'em! Crush 'em! Create 'em! Again and again — in the Time Machine!" The time-lapse TV commercial showed a square plastic lozenge magically melt and transform into an octopus. Reheated, the creatures were loaded into a vise-like press and squeezed back into the original square shape, complete with an embossed Mattel logo. A space-age "shape memory" plastic made it all possible. During manufacture

the molded plastic creatures were irradiated with a high-energy electron beam that cross-linked the polymer's molecules, permanently locking them into shape. These extra connections within the shape could be deformed temporarily into a square brick but when reheated they sprang back into their original creature shape.

The look of the toy was pure 1960s futurismo with a snazzy metallic red housing, shiny zinc fittings, and a transparent transformation chamber with a swiveling door. The assortment of creatures included tiny dinosaurs, kooky spacemen, and mini monsters. The instruction sheet flipped over to make a jungle island backdrop. Even the package's shipping tray was vacuum-formed into a volcanic rock pit. The entire toy was like a sci-fi monster movie set shrunk down into miniature toy form.

The ominous warning stamped on the toy — *Caution: Contact with rivets or plastic parts may cause burns* — was also from another time, one before the Child Safety Protection Act.

One hugely popular 1960s maker toy did have an infamous safety problem. Like the Thingmaker before it, Incredible Edibles used a heated oven (with a hinged cover in the form of a bewigged, buck-toothed bug) to let kids mold squiggly spiders and squirmy worms. The fresh twist was that the finished product was actually edible. The ingredients listed glycerol and tapioca starch, sweetened with sodium cyclamate and saccharin.

The moldable comestible was dubbed Gobble-Degoop and marketed as "sugarless." Parents were repelled more by the taste than by the fun "grossout" theme, but kids gobbled it up. The unforeseen problem was that diabetic kids were sickened by the mixture's starch, which turned to sugar when digested. After $50 million in sales, the FDA allowed Mattel to put warning stickers on all the toys already in stores instead of recalling them.

An earlier candy-making toy promised real sugar in its most kid-appealing form: cotton candy! Commercial machines were expensive and complicated, with spinning electrical coils and strong motors. A kid's only chance for the rare treat was a trip to the circus or state fair. But in 1962 Hasbro's affordable, battery-powered Hokey Pokey Cotton Candy Machine spun real cotton candy at home anytime.

CANDY CRAFTS:
Figs. M and N: Incredible Edibles used an oven to let kids mold squiggly spiders and squirmy worms out of Gobble-Degoop. **Figs. O–Q:** The Hokey Pokey Cotton Candy Machine spun real cotton candy at home.

HASBRO **HoKey PoKey**
Cotton Candy
machine
Safe! OPERATES ON 'D' SIZE BATTERIES

MAKES REAL COTTON CANDY AT HOME... FUN AND EASY TO USE!

A red scalloped plastic base supported a deep aluminum drum. Mom loaded a few spoonfuls of sugar into a metal cup and heated it over a kitchen burner. Using the metal tongs, she lifted the cup of molten sugar into place atop the toy's central shaft and turned on the motor. The molten sugar was flung out of the madly spinning cup in an instant to make one small batch of cotton candy. The metal cup then could be cooled down on the included asbestos stand. Asbestos? Flying molten sugar? The moon-faced kids illustrated on the package happily twirled their paper cones of cotton candy, sweetly oblivious to any potential domestic dangers.

Toys offer kids a way to emulate grownups and play pretend, and maker toys are no different. Dad's basement woodworking "man cave" was the inspiration for Kenner's 1962 Motorized Home Workshop. Although scaled down, the versatile styrene toy could be configured into seven different power tools, including lathe, jigsaw, drill, and sander, just like Dad's Shopsmith.

The gimmick that made it all safe for Junior? Instead of wood, the raw material was colorful, open-cell urethane foam, and the saw and drills were made of plastic. The battery-powered motor "stopped at the press of a finger." It was fun to make clouds of foam "sawdust" with the motorized tools as you cut, sanded, and turned your project on the lathe.

When it comes to mass-producing plastic parts, toy companies use a process called injection molding. A two-part metal mold is held tightly closed in a horizontal hydraulic press as molten thermoplastic is injected at high pressure into the mold cavity. The mold is opened and the cooled plastic part is ejected. Unlike vacuum forming or casting, injection molding with its closed mold creates parts with perfect detail on all sides.

What better maker toy than a miniature version of the very process used to make real toys? Mattel's Injector from 1969 had a plug-in heater and a hand-powered injector piston.

To use it, you just slide open the chamber and insert small plastic pellets (called Plastix). While the plastic melts, choose your mold. A small toggle provides the clamping pressure to hold the mold halves together. Slip the mold under the nozzle and push down firmly on the lever to squirt a piston full of hot

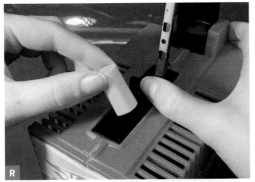

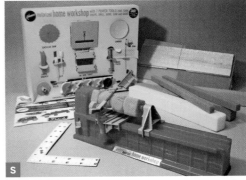

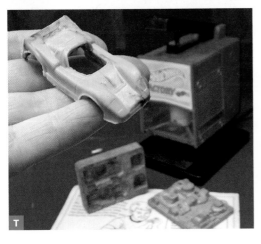

> You can use the clear plastic or metalized squares for any mold.
>
> **IMPORTANT!**
> Metalized plastic goes on the heating chamber blue side up.

DESKTOP GARAGE: Figs. R and S: Kenner's 1962 Motorized Home Workshop promised "planes that fly and boats that float" but the flimsy foam made fuzzy and feeble results. **Fig. T:** Mattel's 1969 Injector had a hand-powered injector piston for molding hot plastic. **Figs. U and V:** The Hasbro Kid and Matty Mattel — these sworn rivals vied to rule the maker toy market.

plastic into the mold. With different themed versions you could mold your own Hot Wheels Factory car bodies or Western World cowboys and Indians.

Depicted on the packages and molded right into the toys were the company mascots. These peppy personalities delivered TV taglines and were even pressed into service enlivening otherwise bland instruction sheets. Matty Mattel gave tips on vacuum forming, the Hasbro Kid shilled the latest Hassenfeld Brothers offerings, and the Kenner Gooney Bird screeched "It's Kenner! It's fun — *Squawk!*"

So where are the maker toys of today? Social trends have changed. Today's overscheduled kids don't have as much free time to spend waiting for long heating and cooling cycles to make a leisurely batch of Creepy Crawlers.

Kids now "grow older younger": whereas toy manufacturers of yore could market toys to kids ages 6–12, today's "tweens" are already using computers and getting their first iPods. For them, the toy box is forgotten; they aspire to create with real, adult materials and tools, so they're at the yarn shop, art store, or Maker Faire, not the toy store. Toymakers

must now focus on younger ages that are inherently less patient, capable, or interested.

And despite clever updates by manufacturers to address modern toy safety standards, it's hard to compete with the high-tech appeal of video games and electronics.

The economy has changed, too: just as manufacturing industries are being replaced by an information-driven service economy, maker toys with their "thing-making" play pattern are giving way to the virtual experiences of electronic games and websites aimed at kids.

Still, the spirit of DIY lives on, and these maker toys engendered an interest and curiosity in many kids who grew up to be the makers of today. What was your favorite maker toy? And what are your favorites today? Tell us at neotoybob@yahoo.com and makezine.com/20/makertoys.

Bob Knetzger is a designer/inventor/musician whose award-winning toys have been featured on *The Tonight Show*, *Nightline*, and *Good Morning America*. He wrote the project "Kitchen Floor Vacuum Former" in MAKE, Volume 11.

Wooden Mini Yacht

THIS AUTHENTICALLY RIGGED MODEL BOAT SAILS ACROSS POOLS AND PONDS.

BY THOMAS MARTIN

When my son was 3 years old, I made a small bathtub boat with him, using scrap wood and a piece of dowel. It lasted much longer and got more of his attention than any dollar-store bath toy, and about six years later we decided to try building a larger boat for the pool and local ponds we fished.

Here's the result of our experimentation: a simple and worthy pond sailer that's rigged and scaled like a real yacht. You can build it in a weekend using readily available materials and tools.

>> Build Your Boat
Time: A Weekend Complexity: Easy

1. Prepare the sailcloth.
It's hard to find waterproof fabric that's easy to cut and won't fray. You can make your own by stretching ripstop nylon loosely over a frame or 2 hangers, and spraying it lightly (in a well-ventilated area) with polyurethane. First spray up and down, and then back and forth, until the fabric is well coated but not saturated. Let dry overnight.

2. Mark and cut the parts.
Download the project plan at makezine.com/20/woodenboat and print it at full size. Following the plan, measure and mark the mast, jib boom, and mainsail boom lengths on the ¼" dowel. Trace the hull from the printed pattern onto the top and 2 ends of the cedar block; cut templates or use carbon paper. Draw the keel and masthead crane patterns on the brass strips, and draw the bowser (rigging clip) pattern 8 times on the thin plastic.

Cut and drill all the parts. Any fine-tooth saw will

Photography by Thomas Martin

cut the dowel, or you can roll it under an X-Acto blade and snap the score. Heavy-duty shears or a hacksaw will cut the brass; be sure to file away the sharp edges afterward. You can saw or file down the hull's shape, then use a hobby knife or thin chisel to excavate the slot for the keel. Drill all holes, plus pilot holes for the screw eyes (in the hull, just poke pilot holes in by hand with a thumbtack).

Finally, file, sand, and smooth all parts. The more time you spend here, the better — especially if you plan to use a clear finish over the wood.

3. Mount the keel.

On the underside of the hull, mask both sides of the keel's slot with tape. Wearing gloves, and in a well-ventilated location, mix and spread some 5-minute epoxy into the slot using a scrap stick or wooden match. Slide the keel into position and hold it there while the epoxy cures (Figure A). You can square it up using a business card on each side. Use a gloved finger to smooth the epoxy along the joint line, and fill any voids with more epoxy.

4. Finish the wood.

Finish the hull uniformly, or for a big-boat look, paint the outside of the hull and stain the deck (Figure B).

Sand the hull with 100-grit paper over a sanding block, and again with 150-grit. Apply a first coat of paint or varnish, and re-sand with 180-grit before each subsequent coat.

For a stained deck, first paint the hull upside down, then re-sand the top perimeter to remove any overspray. Rub stain into the deck and edge, let dry, and coat with varnish or polyurethane.

For the mast and boom pieces, bevel the cut edges for a more finished look, then sand with fine grit to remove any fuzz. Stain if desired, and cover with at least 2 coats of varnish or polyurethane sealer, sanding lightly between coats.

MATERIALS

Western red cedar wood, 2"×4"×18" for the hull. From a lumber yard, or better yet, see if a local fencing contractor can sell or give you an offcut.
¼" hardwood dowel, 36" length
Brass wire brad, #16×1¼" **for the clew hook**
Screw eyes, zinc-plated steel, ½" **(11)**
Brass strip, .032"×½"×12" **for the masthead crane**
Brass strip, .093"×2"×12" **for the keel**
¹⁄₁₆" sheet plastic, about 2" square **cut from the bottom of a milk jug**
Metal eyelets (grommets), ⁵⁄₃₂" (4mm) diameter **(8)**
Ripstop nylon, 2'×2' **for sails**
Braided dacron line, 30lb–80lb test, 6' **for rigging, from a fishing supply store**
Spray paint and/or stain (1+ cans)
Spray polyurethane
5-minute epoxy
Cyanoacrylate ("super") glue or wood glue

TOOLS

Scissors **or carbon paper to avoid cutting the plan**
Needlenose pliers, masking tape, gloves
Small hammer or mallet
Hacksaw
Jigsaw or coping saw
Bastard file or 4" hand rasp and file
Hand drill and bits: ¹⁄₃₂", ¹⁄₁₆", ¹⁄₈", and ¼"
Eyelet tool **such as Dritz #104T**
Sanding block and sandpaper: 100, 150, and 180 grit

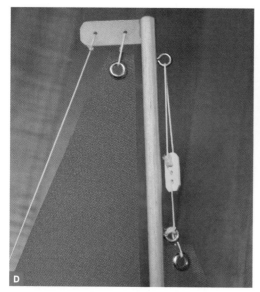

D

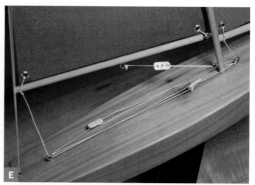

E

5. Assemble the mast and booms.

Cut a slot in the top of the mast and glue in the masthead crane. Once that's secure, follow the plan to install all screw eyes: 4 to the mast, 1 on the fore end of each boom, 1 more on the mainsail boom (for the boom vang), and 4 to the deck. Screw these in until the shank of the screw is completely into the wood. Insert the brass brad down through the hole in the jib boom and bend it into a clew hook (Figure C, previous page).

Use needlenose pliers to open the mainsail boom eye, hook it onto the eye on the mast, and close it. This forms the gooseneck, the joint that lets the boom swing from side to side (Figure C, far left).

Press the mast down into the hole in the deck with the masthead crane centered aftward, and tap it gently down into its hole with a hammer.

6. Add the sails.

After the sail material is dry, trace and cut it to the plan patterns. Lay the boat on its side on a hard surface with the masts and booms in place and fit the sails to the areas for rigging. For the grommets, cut a small X at each sail corner, insert a grommet up through the hole, press the cloth down around it, and tap the grommet flat with the eyelet tool until it firmly grips the cloth.

NOTE: It's a good idea to practice setting grommets first with a couple of sailcloth scraps and extra grommets.

It's time for rigging. Knot and cut a short length of dacron line, thread it through a bowser, and string the boom vang. For these and all other knots, add a *tiny* drop of cyanoacrylate glue immediately after tying; the line is slippery and won't hold knots otherwise.

Use 5" lengths of line to tie each sail grommet to its corresponding screw eyelet or drilled hole with a square knot (Figures C–E). You'll need about 10" for the top of the jib sail, which threads through 2 eyelets before tying off to the uphaul bowser.

Referring to the plans, tie the 4 lower connections on the booms first, and then add the upper lines for tension, so there are no wrinkles in the sails along the booms. Thread a bowser onto the jib uphaul as indicated: for their final tensions, you'll adjust the jib using the uphaul at the top (Figure D), and the mainsail using the boom vang (Figure E).

For the backstay, tie in a long length of line at the masthead crane and install a bowser, routing the line through the eyelet at the stern.

Tighten the backstay and the sails so that they're fairly tight but the mast is not bowed forward or aft.

Finally, add the 2 lines called sheets. For these, cut two 15" lines. Tie each one through the hole in the aft end of a boom, thread it through the sheet eyelet on the deck just underneath, then through 2 holes in a bowser, through the other sheet's eyelet, and finally through the last hole in the bowser, double-knotting the line (Figure E).

NOTE: It's important to tie the bowsers exactly as shown on the plan to make them work.

The sheets let you adjust the angle (trim) of the sails — slack for downwind sailing or tight for crosswind — letting you cross a pond or pool in any direction that isn't too close to directly upwind.

Now it's time to go sailing!

Thomas J. Martin is an artist, illustrator, writer, and blogger at www.tmrcsailplanes.com and www.aerosente.com.

More Kid-Friendly Projects

Volume 05, page 72

Volume 02, page 172

Volume 05, page 78

Volume 10, page 110

Marshmallow Shooter
Volume 02, page 172
BY SAUL GRIFFITH, NICK DRAGOTTA, JOOST BONSEN
Revenge is sweet with your PVC marshmallow shooter.

Backyard Zip Line
Volume 05, page 72
BY DAVE MABE
Be the hit of your hood with a high-flying, tree-to-tree transporter.

Soda Bottle Rocket
Volume 05, page 78
BY STEVE LODEFINK
Build a high-performance rocket powered by air and water.

Tabletop Biosphere
Volume 10, page 110
BY MARTIN JOHN BROWN
Create a sealed system that supplies a fresh-water shrimp with food, oxygen, and waste processing for a desktop journey of 3 months or more. Our "econaut" has been alive for 2½ years!

The Night Lighter 36 Spud Gun
Volume 03, page 108
BY WILLIAM GURSTELLE
Launch potato chunks 200+ yards with this stun-gun-triggered, high-powered cannon with see-thru action.

Vibrobot
Volume 10, page 119
BY MARK FRAUENFELDER
Make a twitchy, bug-like robot out of a toy motor, a mint tin, a coat hanger, and some paperclips.

Styrofoam Plate Speaker
Volume 12, page 131
BY JOSÉ PINO
Get surprisingly good sound from disposable picnicware.

Robotic Hand
Volume 19, page 148
BY NICK DRAGOTTA AND SAUL GRIFFITH
Extend your grasp with a robotic hand made from drinking straws.

Medicine Man Glider
Volume 17, page 108
BY RYAN GROSSWILER
Make a 5-foot-wingspan glider inspired by 1930s stick-and-tissue designs.

Travel-Game Mod
Volume 08, page 17
BY HARRY MILLER
Make a travel edition of your favorite, but house-bound, board game.

Invisible Ink Printer
Volume 16, page 92
BY MIKE GOLEMBEWSKI
Modify an ink cartridge to print with invisible lemon juice.

Compressed Air Rocket
Volume 15, page 102
BY RICK SCHERTLE
Send this 25-cent rocket hundreds of feet in the air.

Archery Bows
Volume 16, page 169
BY DAN ALBERT
Make an archer's bow from eco-friendly, fast-growing wood.

Household Worm Bin
Volume 18, page 78,
BY CELINE RICH-DARLEY
Let hungry, squirmy wigglers take out the household trash.

Boing Box
Volume 12, page 116
BY MARK FRAUENFELDER
Build a fun, one-stringed twangy instrument.

Gauss Rifle
Volume 01, page 12
BY SIMON QUELLEN FIELD
Build a linear accelerator for studying high-energy physics.

Wizard Crackers
Volume 04, page 179
BY CHARLES NEVEU
Make trick crackers that go off like a gunshot and burst into a ball of flame.

10-Rail Model Rocket Mega-Launcher

MAKE THE CUB SCOUT ROCKET DERBY A BLAST.

BY DOUGLAS DESROCHERS

At last year's annual Cub Scout Pack 1346 rocket derby, nearly 100 rockets were launched, testing the attention span of many younger scouts. For this year's event I wanted to build a system that would create more excitement and keep the pace of the launches moving along. This 10-pad mega-launcher is the result of that initial inspiration.

This system gives kids the fun of pressing their own launch buttons, and for added drama, sounds a klaxon before each launch. For the finale, a "Mega Launch" switch shoots off all 10 rockets at once.

I also wanted to minimize the chance of an unfired rocket left on the pad while the others soared skyward, disappointing a child. Therefore, this system has always-on igniter continuity checks, using LEDs to show which igniters are ready. Blocking diodes allow dual-use of the wire harness — for continuity check and launch voltage — thus halving the number of wires to the launch bar.

The system also needed to withstand rough handling from lots of kids, so I used copious amounts of hot glue, heat-shrink tubing, and zip ties. And along the way, another goal became showing the kids that a system like this isn't complicated, which is why I put the workings behind clear plastic.

Photograph by Jason Hornick

MATERIALS

Lexan panel, 0.220" thick, 21"×36" **A 30"×36" sheet is about $40 from Home Depot.**

Perf board, about 3"×4"

Lego klaxon alarm sound brick **item #55206c03 from BrickLink (bricklink.com)**

Toggle switches with red safety covers (2) **I used MPJA (mpja.com) parts #12219 and #16100.**

SPDT key switch **for the Power switch**

7805 5V DC voltage regulators (2) **All Electronics (allelectronics.com) part #7805T**

555 timer IC chip **All Electronics #LM555**

10kΩ resistor

Yellow LED and 150Ω resistor, ¼W or ½W (optional) **for an Arm switch indicator light**

100kΩ potentiometer **All Electronics #POS-104A**

47μF capacitor

Quick-connect wire connectors with headers: 4-pin (3), 10-pin (1) **All Electronics #CON-244 and #CON-2410**

Angle brackets (6)

10-wire, 22-gauge bundle cable, 25' **for the launch bar firing cable; All Electronics #10CS22**

14-gauge insulated wire, 30' **for the launch bar grounding cable**

22-gauge hookup wire, 150'

Large 12V battery **I used a lawn tractor battery from a discount store, about $20.**

Lamp cord **for connecting battery to circuit**

Spring clamps (2) **for connecting battery to circuit**

2×6 (1½"×5½") pressure-treated lumber, 8'

2×3 (1½"×2½") lumber, 12'

¾" plywood scrap, about 2' square

13mm bolts, at least 2" long, with matching wing nuts (2)

Cabinet hinges (2)

Construction adhesive

Typical workshop stuff: epoxy, screws, hot glue, zip ties, heat-shrink tubing, bolts

FOR THE 10 LAUNCH PADS:

2-outlet electrical box cover plates, steel (10) **from hardware or home improvement stores**

Metal rods, ⅛"×3' (10) **from a hardware store**

Momentary push-button switches (10) **to fit the Lexan panel**

1N4004 blocking diodes (10)

LEDs, ultra bright, clear lens (10) **All Electronics #LED-94**

510Ω resistors, ½W (10)

Mini alligator clip jumpers (10) **MPJA #16434**

TOOLS

Soldering iron

Table saw

Drill press and bits, including #9, ⅛", and Forstner bits

Hot glue gun

Wire strippers/cutters

Multimeter

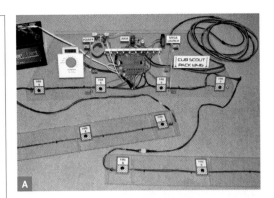

A

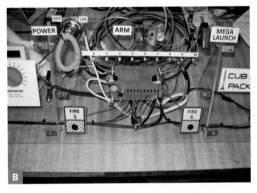

B

▶▶ Build Your Mega-Launcher

Time: **1–2 Weekends** Complexity: **Moderate**

Layout

The mega-launcher consists of a 10-rocket launch bar and its control station connected a safe distance away by a 25' cable. The control station fits on 2 long folding tables arranged end to end, with firing stations 1–3 on the left, 8–10 on the right, and 4–7 on the control console in the middle (Figure A).

The console (Figure B) carries the system-wide Power, Arm, and Mega-Launch switches. For safety, the main power switch on the control console is a key switch; this lets you pull the key to prevent overly excited scouts from firing rockets while someone is still working at the launch bar.

Turn on the power, and a row of 10 continuity LEDs tells you which launch pads have igniters in place that are ready to fire. Flip up the Arm switch and the klaxon sounds, signaling that the firing station and Mega Launch buttons are now enabled!

Circuit Design

The system is powered by a 12V battery, and continuity to each igniter is ensured by an LED that's

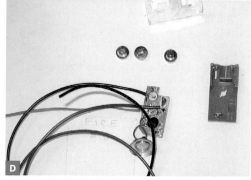

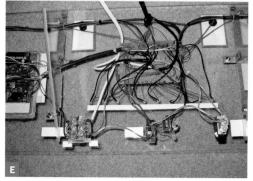

bright enough to see in daylight. To accomplish this, the circuit constantly trickles 5V through each igniter in series with the LED and a 510Ω resistor. Pressing a firing button overrides this series and delivers full voltage to the igniter.

This is the neat part of the system; one DC wire running to the launch bar serves as both the continuity check in one direction, and provides firing 12V DC current in the other direction when the firing button is pressed. See makezine.com/20/ megalauncher for the circuit schematic. Naturally, I went through a number of test igniters before I found component values that would keep the LED lit but did not spark the igniter.

I included the alarm, which sounds when the system is armed and the rockets are ready to fire, for both fun-factor and to give adult supervisors some warning to get everybody away from the launch pads. The Lego sound brick I used only sounds 5 klaxon "beeps" per triggering, so I added a pulsed trigger circuit based on a 555 IC chip to extend the duration of the alarm and make it adjustable with a 100K potentiometer. I also added a yellow LED and 150Ω resistor that indicate when the 555 chip is triggering.

The circuit has two 5V DC voltage regulators

because the peak draw of the continuity LEDs is 100mA and the regulators are rated to 1A. With a 10-pad system, this means dedicating one regulator to the LEDs and using the other for the alarm.

Electronics

I built most of the control circuitry on a piece of perf board, separating the components to allow plenty of space in back for the power and ground connections. I mounted the circuit board just behind the Lexan control panel using 1½" pan head screws, so you can see all the components up close but they still have breathing room.

I mounted the switches, LEDs, and alarm speaker in the Lexan panel, drilling LED holes with a #9 bit and larger holes with a Forstner bit. The Arm switch is a double-pole switch, and I used a military aircraft weapons switch for the Mega Launch function, but any SPST will work.

I arranged the 510Ω resistors that lead to the continuity LEDs along the top edge of the circuit board, allowing short connections to the row of LEDs in the panel just above (Figure C). To connect the alarm, I disassembled the original Lego brick, soldered leads to the mini board inside for trigger voltage and ground, and spliced in 2 more wires

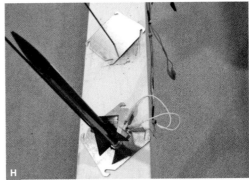

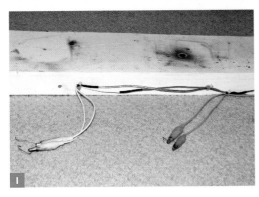

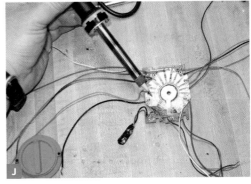

to bring the speaker off-board and let it reach the panel (Figure D).

Panels and Wiring

The central control panel is built into a 36"×11" Lexan panel with two 9"×6" triangles cut from the top corners and reattached with angle brackets to make a standing base (Figure E). The 2 side firing panels were built into 5"×30" Lexan strips, each connected to the console by a 4-pin harness. I used small zip ties to bundle wires together throughout. Be sure to check the switch wiring with a voltmeter prior to soldering.

Power for the system comes from a big 12V battery that sits behind the control panel. With a full charge, this 230-cold-cranking-amps battery lasts well past 200 ignitions. To connect it to the circuit, I used lamp cord and medium-sized clamps to grab the battery terminals.

For easier storage, I used a 10-wire connector for detaching the cable to the launch bar, and two 4-pin connectors (3 launch, 1 ground) for disconnecting the wire bundles leading to each side launch panel. The 14-gauge launch bar grounding cable also detaches via a 4-pin connector, in order to prevent a current bottleneck at the disconnect point.

Connecting four 22-gauge ground wires, all soldered together in parallel to the heavy grounding wire, splits the load (Figure F).

Launch Bar

The launch bar is straightforward. To elevate it, I built 2 foldable sawhorse legs using 2×3 lumber, a hinge, and a triangular piece of plywood. One side of the triangle is fixed to one leg with wood screws, and the other is drilled out for a removable nail, so the legs can be folded for storage (Figure G).

The launch bar itself is an 8-foot 2×6 board with 10 individual pads consisting of a metal outlet plate for the blast shield and a ⅛"×3' steel rod for the launch rod. I used a drill press to set ⅛" holes 9" apart on the board for each rod, and drilled ³⁄₁₆" holes through the plates to let the rods pass through. You could use construction adhesive to glue the rods into the holes and glue down the plates (Figure H), but I left them unglued for easy storage and transport.

To allow pivoting of the launch bar for wind adjustments, I cut the heads off 2 large 13mm bolts, then drilled the launch board ends and glued in the bolts. The bolts pass through holes at the top of each sawhorse triangle, secured with wing nuts on the other side.

Each station has a pair of mini alligator clips to clip onto the igniters, 1 for voltage and 1 for ground. To wire the voltage clips to the 10-wire cable in the middle of the bar, I ran a 22-gauge red wire from each station to a pin on the 10-pin connector. I used lots of heat-shrink wrap and zip ties, but after 210+ rocket launches, still wound up with 2 broken wires.

Wiring the igniters' ground sides was a little more complicated. At the ground cable's 4-pin connector, I soldered 4 strands of 14-gauge speaker wire, and then ran a short and a long strand to the left, and a short and a long strand to the right. I attached the alligator clips by soldering to the ends of the speaker wires where possible, or else soldering mid-wire after scraping away some insulation, then covering with heat-shrink (Figure I, previous page).

Design Overkill

Eagle-eyed readers will note in the photographs an unexplained feature that falls into the overkill category. I hacked a multimeter to make a knob for the control console that you can turn to check that 12V DC is running across each of the 10 blocking diodes.

Long story short, I soldered jumpers to the multimeter board to keep it permanently at its 20V DC setting. Then I carefully cut and arranged copper tape traces on a thin plastic sheet to match the contact points of the multimeter's knob (Figure J). The

traces connected to a wire bundle that tapped into the blocking diodes for each igniter. Turn the knob, and you can check each one.

This all took as much time to get right as the rest of the project, and after 180+ launches I realized that it represented almost no value added. I thought that seeing the actual voltage at each pad would help troubleshoot, but the continuity LEDs worked perfectly on their own, and saved many launches where there was a short circuit at the pad. You can also just use a multimeter to check the voltage across the battery before each launch.

Lessons Learned

The mega-launcher made our Cub Scout Rocket Derby a blast. More than 120 rockets were fired in less than 90 minutes — a huge improvement over single-launch systems! And the klaxon sound was a big hit.

One lesson learned from the test launches is that with the close spacing of the rockets, about 9", an adjacent liftoff could knock an igniter loose or cause the clips to short out. A few pieces of masking tape solved this problem.

Another thing we found: with so many rockets in flight at once (Figures K and L), it was difficult to track all of them through touchdown!

➕ For the circuit schematic, plus photos and a video of the 10-rocket mega-launcher in action, visit makezine.com/20/megalauncher.

Doug "Beads" Desrochers (beads27@cox.net) has been voiding warranties ever since middle school. He graduated from Georgia Tech with a B.S. in aerospace engineering, and served in the U.S. Navy as a test pilot and instructor test pilot as well as on several operational tours. He's currently a civilian systems engineer and test pilot supporting the Department of Defense in Washington, D.C.

Photography by Jason Hornick

The UnaBox

MAKE A KEYLESS, SECRET, WOODEN COMPARTMENT.

BY CLAUDIO BERNARDINI

My architect friends Lorenzo Bini and Roberta Pezzulla made this little project for a very young and dynamic client: Lorenzo's 6-month-old goddaughter. They wanted to make a box that the child could use her whole life to store small, personal, precious things. So they created the UnaBox, taking inspiration from an Indian box (by an anonymous designer) that had an obscure opening system with 2 pivoting lids.

The box's dimensions are based on the golden ratio, and it allows access only to patient and curious people; many adults have tried to open it and failed. Here's how to build one yourself!

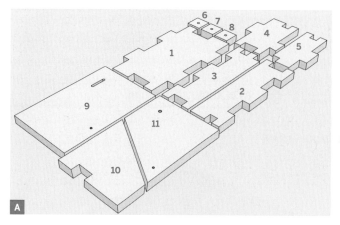

A

B

MATERIALS

Hardwood board 27.5cm×57.5cm×1.5cm thick
Our original box used the best Italian olive wood. Sourcing the wood locally avoids hardwoods with uncertain (and possibly destructive) origins.
Thin wooden peg 3mm×20mm **for the locking pin**
Screw, 4mm×60mm **for the cover pivot**
Wood glue
Natural wax

TOOLS

Band saw or small handsaw
Carpenter's square **ruled**
Pencil
Drill press or drill
Clamps
File
Sanding block with sandpaper, various grits

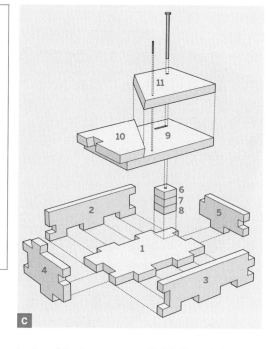

C

▶▶ Build Your UnaBox
Time: **1-2 Days** Complexity: **Easy**

1. Cut the pieces.

The box is made from 11 pieces cut from a single board, plus a screw and a thin peg for the pivot and locking pin. Visit makezine.com/20/unabox for CAD drawings and an assembly sequence (Figures A and C).

To make the pieces, you can send the drawing to a CAD/CAM company and wait for the mailman, give it to a good carpenter, or cut the pieces yourself using a small handsaw or band saw (Figure B). Finish everything roughly with a wood file, and then more precisely with plenty of sandpaper.

2. Assemble the lid and box.

Following the plan, drill a hole in piece 11 and a slot

in piece 9 that your screw will slide through (to cut the slot, drill each end and then thread in a saw blade). Also drill smaller holes in the pieces for the locking pin, running the drill only partway through piece 9. Drill both holes with the 2 pieces stacked together to make sure they align.

Figure C shows how the parts of the UnaBox fit together. With precise cutting, the bottom and sides of the box, parts 1–5, should assemble with almost no glue; use small clamps to hold them while the glue cures (Figure D). Glue stacked pieces 6–8 together and glue them to piece 1, the bottom of the box. Also glue together pieces 9 and 10 for the box's top. Once you're done with the glue and the clamps,

Photography and CAD drawings by Lorenzo Bini, GroS, and Roberta Pezzulla

drill pieces 6–8 to hold the bottom of the pivot screw (Figure E).

Finally, finish the box with a good natural wax and polish it. This makes the most of the wood, and gives it a nice smell and touch that kids love.

3. Learn the secret.

The closed box gives few clues to how it's opened. You need to remove the peg and swing part of the top away before the rest of the top has enough clearance to slide back and swing open (Figures F–I).

So far, we've built 2 of these boxes: the first one with olive wood, the second with cherry. If you build your own UnaBox, the architects would love to hear from you at mail@studiometrico.com.

➕ Get SketchUp and AutoCAD models of the Una-Box pieces, and drawings of the assembly sequence in Flash and SVG, at makezine.com/20/unabox.

Claudio Bernardini lives in Milan, Italy, where he runs the clothing design and distribution firm Comvert. In his spare time he rides a skateboard, writes on blog.bastard.it, and builds things.

THE LOST ART OF LASHING!

STORY & ART: GEVER TULLEY
PHOTOS: JULIE SPIEGLER & SAMANTHA GOUGH

OUR CIVILIZATION WAS BUILT ON A TECHNOLOGY SO *ADVANCED* WE STILL DON'T KNOW EVERYTHING IT'S GOOD FOR.

BUT SOMEWHERE ALONG THE WAY, MOST OF US SEEM TO HAVE *FORGOTTEN* HOW TO TIE THINGS TOGETHER.*

*GEORGE DYSON POINTED THIS OUT BACK IN VOLUME 09. —EDS.

YET, IF YOU CAN TIE THINGS TOGETHER SECURELY, YOU CAN MAKE *ALMOST ANYTHING* FROM PRACTICALLY NOTHING.

CONSIDER THE GOLF BALL — UNDERNEATH THOSE DIMPLES, AN INCREDIBLY LONG RUBBER BAND SUBJECTS THE CORE TO ALMOST *10,000 POUNDS OF PRESSURE.*

THE *SECRET* IS IN THE WRAPPING — *EVERY TURN ADDS PRESSURE.*

WHICH IS THE SAME PRINCIPLE AT WORK IN A LASHING — THE RESULT IS *VERY STRONG*, BUT WITH A *LITTLE BIT* OF GIVE.

Photograph by (middle right) Captain Edward Augustus Inglefield, National Maritime Museum, Greenwich, London

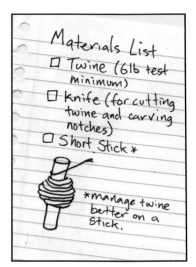

Materials List
☐ Twine (6lb test minimum)
☐ Knife (for cutting twine and carving notches)
☐ Short Stick *

*manage twine better on a stick.

OK, GET YOUR STICKS AND LINE AND WE'LL TIE OUR FIRST LASHING. RELAX —

THIS TAKES A LITTLE BIT OF PRACTICE, SO EXPECT TO START OVER A COUPLE OF TIMES.

ALWAYS KEEP THE LINE TIGHT AND TIDY AS YOU LAY IT DOWN.

KEEP GOING AROUND UNTIL YOU HAVE ENOUGH WRAPS.*

*8 TO 10 FOR TWINE AND STRING.

THEN, TO REALLY CRANK UP THE TENSION, ADD CROSSING WRAPS.

THEN TIE IT OFF WITH A SURGEON'S KNOT (WHICH IS JUST AN OVERHAND WITH AN EXTRA TWIST).

ALMOST DONE — NOW WE LOCK THE SURGEON'S KNOT IN WITH A SQUARE KNOT. THERE ARE FANCIER WAYS TO DO IT, BUT THIS IS SUFFICIENT AND EASY TO REMEMBER.

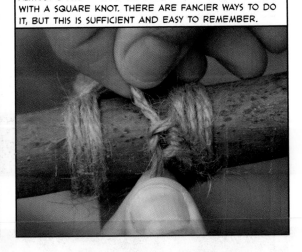

IT'S *CRITICAL* THAT THE LASHING IS TIGHT...

... SO IF ONE GETS LOOSE, YOU CAN WHITTLE A WEDGE *AND POUND IT UNDER THE LINE* TO TIGHTEN IT.

IF YOU NEED TO MAKE A *LONGER* STICK FROM 2 SHORT ONES, OVERLAP AND LASH 'EM.

ADD A FEW LOOPS ACROSS THE LASHING TO TIGHTEN IT UP NICELY.

TYING 3 STICKS CAN BE *TRICKY*, SO INSTEAD OF LASHING ALL 3 TOGETHER ...

... YOU CAN JUST LASH THEM IN PAIRS.

YOU CAN RUN THE LASHING BETWEEN MANY STICKS AT ONCE TO CREATE A *QUICK DECKING*.

A SECURE, FLEXIBLE CONNECTION IS MADE BY RUNNING THE LASHING BETWEEN THE STICKS.

THE SAME TECHNIQUE CAN BE EXTENDED TO ACCOMMODATE 3 STICKS FOR A *TRIPOD*.

THESE TRIPODS CAN BE VERY STABLE, REMARKABLY STRONG, AND CAN BE USED AS A FOUNDATION FOR OTHER STRUCTURES.

SOME DURABLE LINE IS USED TO KEEP THE LEGS FROM SPLAYING.

A NATURAL FORK CAN ADD *STRENGTH* TO YOUR CONSTRUCTION AND YOU ONLY HAVE TO LASH TO THE STRONGEST SIDE — *GRAVITY DOES THE REST.*

IF THE LINE SLIPS ON THE WOOD, CARVE OUT A COUPLE OF NOTCHES TO GIVE IT SOME PURCHASE.

IT'S NOT *EXACTLY* LASHING, BUT WHEN YOU NEED *REAL COMPRESSION*, A FEW LONG LOOPS AROUND A PAIR OF POLES ...

... AND A SHORT STICK MAKE FOR A HIGHLY EFFECTIVE *TURNBUCKLE.*

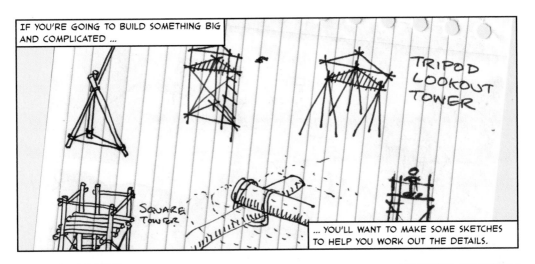

IF YOU'RE GOING TO BUILD SOMETHING BIG AND COMPLICATED ...

TRIPOD LOOKOUT TOWER

SQUARE TOWER

... YOU'LL WANT TO MAKE SOME SKETCHES TO HELP YOU WORK OUT THE DETAILS.

IS *THIS* TALL ENOUGH, MORI?

TALLER!

NOT *EVERYTHING* HAS TO BE STICKS. WE'LL JUST FINISH IT OFF WITH A HEAVIER LINE HERE TO MAKE A RAILING.

STABLE *AND* STRONG — YOU CAN MAKE *ALMOST ANYTHING* WITH LASHING.

DO YOU SEE ANYTHING, MORI?

PIRATES!

GEVER TEACHES MORE THAN JUST LASHING AT TINKERINGSCHOOL.COM.

FIN.

The Weight-Powered Vehicle (WPV)

HOW FAR CAN A CAR GO ON ONE FOOT-POUND OF ENERGY?

BY RICHARD B. GRAEBER

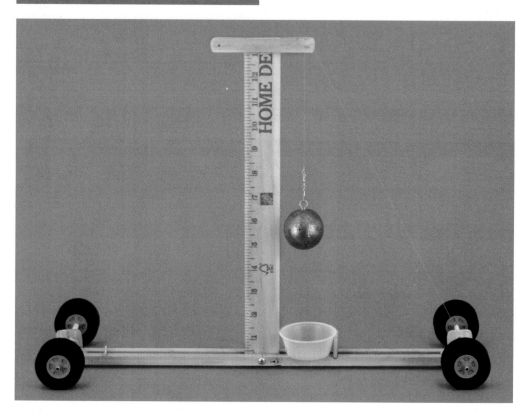

After my son won his high school science class mousetrap-powered vehicle (MPV) competition, I asked him if he knew what a foot-pound of energy was. He gave a vague answer that didn't really nail it. Admittedly, a foot-pound (ft-lb) is an abstract concept — where do you get a pound of feet from anyway?

To illustrate what a foot-pound of energy is and what it can do, I designed a vehicle that derives its propulsion from a falling weight (rather than an unwinding spring as used in an MPV).

This small car, which I call a weight-powered vehicle (WPV), demonstrates the conversion of potential energy into kinetic energy in one of its most rudimentary forms, leading to a better understanding of energy conversion and loss.

Construction

My son and I built the car from pinewood we picked up at the hardware store. We also bought a wooden yardstick, and we found lead sinkers and monofilament fishing line at a sporting goods store. The local hobby shop supplied us with wheels, a small pulley, aluminum tubing for axles, and polypropylene tubing lubricated with teflon powder for bearings. We used these parts, along with a few other odds and ends, to cobble together our experimental vehicle.

Photograph by Ed Troxell

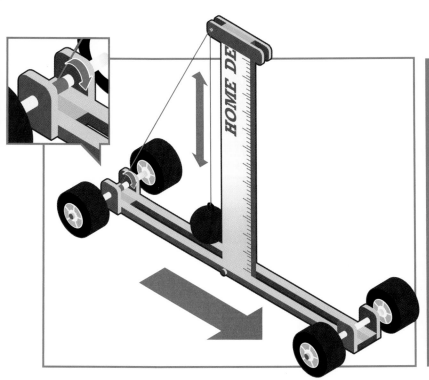

The central feature of the vehicle is a 1'-high mast made from the yardstick. At the top is a pulley, over which the fishing line is wrapped, bearing the 1lb sinker. This drive mechanism is similar to that of an MPV car, but its string-wrapped axle is propelled by a falling mass instead of a spring. In addition, the drive axle has a collar (aka drive hub) to give more leverage to the string when conveying the 1lb force.

Testing and Results

We tested the vehicle with 3 different wheel diameters (2", 2¼", and 3"), 2 drive hub diameters (⅛" and ¼"), and 3 sinkers (6, 10, and 16oz). The vehicle itself, without the sinker, weighed 4oz. The values that worked best for our vehicle design were 2¼"-diameter wheels and a ¼"-diameter drive hub.

When tested with the minimum weight, 6oz of lead, the vehicle failed to move even when pushed. This we attributed to friction and drag. With the 10oz weight, the vehicle rolled from a standstill without assistance, covering 10' in 9 seconds. When the 16oz weight was put to the test, the vehicle spanned the same 10' distance in 12 seconds. We attributed this slower speed to increased rolling resistance of the soft sponge-rubber tires, caused by the increase in vehicle gross weight.

Conclusions

» Our proof-of-concept WPV used 1 ft-lb of energy to travel 10'. As we experiment, we're confident we can build WPV vehicles that can travel much greater distances using the same amount of energy.

» Within a narrow range, gross vehicle weight didn't affect elapsed distance, because very little coasting occurred. The only thing affected was elapsed time.

» Because the WPV uses less force than the MPV, it requires more precision in its construction. Bearing designs that can be used with little performance degradation in an MPV are too crude and apply too much drag in the WPV. The difficulty we experienced in trying to minimize rolling resistance and power-train drag was tougher than originally envisioned, because of the subtle forces involved.

» Overcoming this challenge is ideal for a contest of ingenuity and skill.

Phase II: Energy Recovery

If converting potential energy into kinetic energy gets boring, Phase II could be an "energy recovery of applied force" competition. Within a limited distance, the vehicle that covers the greatest distance — and recovers the greatest amount of kinetic energy when the brakes are applied — wins.

Richard B. Graeber and his son won Saratoga High School's mousetrap-powered vehicle competition and then evolved the concept to the next level.

Teachers' Pet Projects

WE ASKED TEACHERS TO SHARE THEIR FAVORITE CLASSROOM & SCIENCE FAIR PROJECTS.

COMPILED BY LAURA COCHRANE

Candy Dispenser

Time Travel Helmet

Light Energy

Daniela Steinsapir, San Francisco, Calif.

» **Candy Dispenser**
A 7th grader made this electromechanical sculpture during electronic art class. It's a candy dispenser made out of cans. You put the candy on top, it goes through a plastic tube, and by pushing a button, it activates the servomotor, dispensing the candy.

» **The Future of Mankind — Through a Helmet**
Two 7th graders made this during the space arts class. It's a time travel helmet, made out of recycling materials and discarded electronics, and has a video screen inside. The viewer is invited to time travel to the past and the future by wearing the helmet. The video that plays narrates the students' theories about space, technology, and the future.

Vickie Connally, Loving, N.M.

» **Reflected Light Energy**
My favorite science project is an analysis of energy generated from reflected light. Students use solar panels and microamp meters to measure moon-light, comparing phases of the moon, cloudy nights, etc. Power is generated! This leads to great discussions about reflected light.

» **Recycled Playground Equipment**
One year we needed playground equipment at a middle school. The challenge was to build it from recycled materials. We got fantastic benches from old headboards, some cool stuff from large tires, and some neat seating from old tractor seats.

» **Cell Model from Trash**
Use clear plastic food containers or clear plastic bags for the cell membrane. Then chewed gum or other trashed items are molded to represent all the parts. science-ideas.com/3d-plant-cell-model

Rick Schertle, San Jose, Calif.

» **Ice Cream in a Bag**
Ice and salt in one zip-lock bag and cream mixture in another bag, mixed by tossing back and forth, is a classic treat for all ages! My colleague took his 8th graders over to the kindergarten class next door and they made the sweet treat together.
makezine.com/go/ice_cream

Eric Muhs, Seattle, Wash.

» **Pinhole Cameras**
Any small box can be made light-tight, and a simple darkroom can be set up in a corner of the basement or a small bathroom. The creative possibilities are endless: long exposures, double exposures, panoramas, curved pictures. makezine.com/go/pinhole

Earthquake Simulator

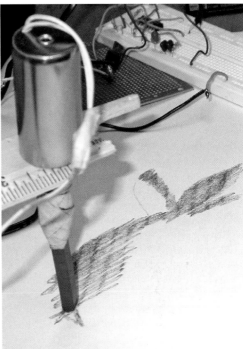

Robot Drawing Machine

» Mobiles

Using related objects, make a mobile a la Alexander Calder that appears to defy ideas about balance. Mobiles teach about rotation, center of mass, and symmetry, and they connect physics and art. Make a large mobile for a public space — we made one using books for our library. makezine.com/go/mobile

» Crop Circles

In a geometry class, we were tired of step-by-step constructions (draw a ray that connects point A on the circle to point X on the line ...). So we started to research and draw original crop circles. Students really liked the "mystery" of crop circles, and enjoyed explaining their processes. We had nice software called The Geometer's Sketchpad, but it was really just a fancy compass and straightedge.

» Musical Instruments

Students define themselves by the music they like, but not many have any relationship to creating music. Simple instruments are a challenge, and you can add structure to a project to make it a richer experience.

Students can record the sound waves with fairly inexpensive software, learn how to play a song individually and with other instruments, and even compose. makezine.com/go/instrument

» Trebuchets

I've been doing this with students for 15 years, and have a lot of ideas about how to get your students to build successful trebuchets. We use simulations, and build our own computer models and scale models before committing to the larger investment of a serious hurler. makezine.com/go/trebuchet

Tom Zimmerman, San Jose, Calif.

» Water Rocket Engine Test

Water rockets use a mixture of air pressure and water for propulsion. For a science fair project, my students attach a rocket to a wheel and measure how long it spins to determine the most powerful mixture of air pressure and water. (They actually used a mixture of hydrogen and oxygen and water, but plain air and water is simpler and safer.) [See "Hydrogen Oxygen Bottle Rocket" on page 90.]

» Robotic Drawing Machine

Two students wanted to make a robot that draws, but they had no programming experience, so we devised a contraption that used the ancient pantograph with a light sensor, solenoid, and piece of charcoal to make a mechanical image reproduction system. makezine.com/go/robotic_drawing

» Earthquake Simulator
An off-center weight is attached to a drill. The drill is attached to a piece of wood suspended in a box. When the drill turns the off-center weight, the wood platforms shakes, simulating an earthquake.

Here's a manually controlled Shaker Table: makezine.com/go/earthquake (PDF), and others here: makezine.com/go/shaker_tables.

Steve Davee, Portland, Ore.
» Portable Emergency Shelter
My 5th-grade math students created an inexpensive, portable emergency shelter, using three sheets of 4'×8' plastic corrugated plastic. They decided to design with preschoolers in mind, so they measured and averaged the volumes of our youngest students.

After initial brainstorming and designs in their Maker's Notebooks, they made scale models of shelters (and preschoolers) out of paper to test and refine their ideas. Once they agreed on a final design, they constructed a shelter that was a successful blend of efficiency, comfort, portability, and utility.

Jane Gerughty, Pacifica, Calif.
» Plastic Projects
I like to shrink #6 plastic fast-food containers. Students draw something science-related with Sharpies, punch holes for hanging, and then shrink the science art in a toaster oven.

I do a mini unit on the history and use of plastics and environmental concerns. I show them a small plastic preform for a water bottle and a large pre-form for a larger soda bottle to emphasize change of density and relate it to mass.
» Owl Pellet Dissections
I link this lesson to food chains. I have students iden-tify the bones and re-create skeletons [of the animals the owl ate]. makezine.com/go/owl_pellets (PDF)
» Strawberry DNA
Another favorite is DNA extraction from strawberries. It's a nice way to see DNA. makezine.com/go/dna

Alicia Hardy, Oakland, Calif.
» Roller Coaster Design Project
Using inexpensive materials, you take foam pipe insulation and cut it in half, making two sets of semicircular tracks. All you need is a marble to begin an exciting challenge.

You can create distance challenges or have students try to get the marble into a cup several meters away. You can even have a "make the biggest loop-de-loop" contest. Students get to express their creativity and explore the physics of energy, all without even knowing it!
makezine.com/go/roller_coaster

Richard Delwiche, San Francisco, Calif.
» pH Indicator
Rather than using expensive and toxic solutions, I prefer to have students make their own. The simplest and most widely known is cabbage juice, obtained by either squeezing or boiling.

Another awesome indicator is turmeric. I put powdered turmeric in alcohol, then soak up the yellow solution into coffee filter paper.

After it dries, strips of this paper can be used to indicate the presence of a base. I hold it over the mouth of an ammonia bottle to show that even the vapors turn it to a vivid red.
makezine.com/go/ph (PDF)
» Membrane Aerophones
These are musical instruments that sound a little like oboes. You blow into a straw to pressurize a small chamber made from a film canister covered with a membrane. As air pushes up the membrane it escapes the chamber through a PVC pipe. The tube has finger holes that can be covered like a recorder to play a scale. makezine.com/go/aerophone (PDF)
» Oil Drop Photometer
We have also made simple and elegant "oil drop photometers." makezine.com/go/photometer

Gever Tulley, Montara, Calif.
» Wind-Up Juice-Bottle Boats
Every kid gets two empty 15-ounce juice bottles, a couple of rubber bands, and some sculpture wire. At first it seems impossible, but then it's clear that the possibilities are endless.

Aero- and hydrodynamics are explored, and friction and tension are diagnosed and managed.
makezine.com/go/bottle_boat
» Deconstruct and Reanimate
Obsolete storage devices such as cassette tapes and floppy drives are dismantled and the parts are probed with AA batteries to see what happens.

Electromechanical principles are discovered, and basic concepts of circuits and polarity are explored.
» Potential Energy
A few hex nuts, some string, and a couple of coat hangers are all that's needed to play with potential energy. Wire sculptures are animated with falling weights and unspooling line. Simple mechanical linkages are invented and refined, and a deeper understanding of gravity may be revealed.

Portable Emergency Shelter

Juice-Bottle Boats

Chain Reaction Contraption

Cardboard Automata

Mendocino Motors

Photography by (clockwise from top left) Nicole Simpson, Julie Spiegler, Chris Connors, Karen Wilkinson, and Luigi Anzivino

Mike Petrich, San Francisco, Calif.

» Chain Reaction Contraption

One of my favorite projects is to build a collaborative Chain Reaction Contraption, inspired by Rube Goldberg or Heath Robinson (the U.K. version of Goldberg). *Pythagoras Switch* is another inspiring resource, a Japanese TV show highlighting short chain reactions.

We often impose a theme for the construction; we've created chain reaction contraptions based on Einstein, Pi/e (3.14159 and dessert), and Love.

Students encounter high-level design and problem solving. Experiments with completing circuits, potential and kinetic energy, rotational motion, leverage, humor, metaphor, and collaboration are all required to build and contribute to the event.

» Cardboard Automata

Inspired by the clever Cabaret Mechanical Theatre in the U.K., we host an activity that allows teens to construct their own narrative, kinetic machines, utilizing cams, gears, levers, and other simple mechanisms. Unlike other science activities that use art to "decorate" the experiment, this automata activity combines science, art, and storytelling in equal parts. It is challenging, incorporates common materials in unusual ways, and allows for individual expression. makezine.com/go/automata (PDF)

Chris Connors, Duxbury, Mass.

» Mendocino Motors

One project I've done is the Mendocino motor. It's a solar-powered, magnetically levitating motor. Here's a document that lays out a lot of the process: makezine.com/go/mendomotor

» Vibrobot Variations

I also like variations on vibrobots (as seen in MAKE, Volume 10). A fun project we have done is to disassemble CD players, make battery holders, and then turn them into vibrobots.

With a limited set of tools, a couple of discarded drives, and a few hours' time, this project introduces students to variations of manufacturing design, sourcing supplies for future projects, and the joy of discovery inside previously mysterious devices. makezine.com/go/vibrobot

ADDITIONAL RESOURCES:

Museum of Science and Industry activities
msichicago.org/online-science/activities
Exploratorium Science Snacks
exploratorium.edu/snacks
List of organizations that promote hands-on learning: makezine.com/20/teachers

Marble Adding Machine

MAKE A MECHANICAL, GRAVITY-POWERED, BINARY CALCULATOR THAT USES WOODEN LEVERS AND CHANNELS TO COUNT.

BY MATTHIAS WANDEL

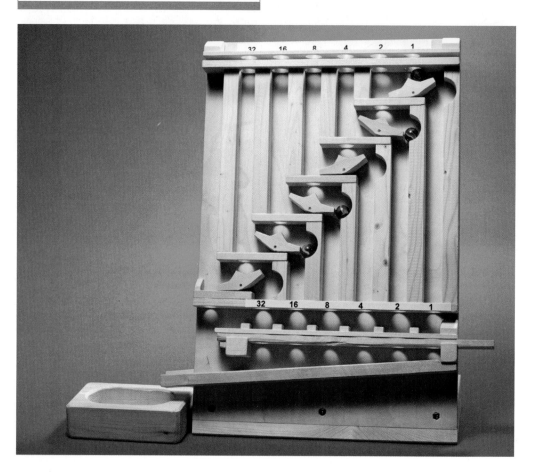

Computers add binary numbers constantly, but we never see how. This elegant machine does the math using glass marbles.

I started building marble track machines years ago using Lego. I experimented with all sorts of crazy ways for the marbles to descend. One was a rocker that shunted a stream of dropping marbles one-by-one to alternating sides. If you cascade three of these toggles down to the left, every marble flips the rightmost toggle, every second marble flips the middle toggle, and every fourth flips the leftmost. Interpret each toggle's state as left = 0 and

right = 1, and you have a binary counter. Add more toggles, and it can count more.

I realized I could turn the counter into an adder by dropping marbles onto toggles other than the rightmost. I added a marble hold-and-release shelf up top to act as an input buffer, another underneath as an output buffer, and a clear-register mechanism to reset all toggles. I moved from Lego to wood, and refined the design a few times. Here's the latest edition.

How It All Adds Up

The adder uses marbles (and the absence of marbles) to represent bits. The input holder holds one binary number while the toggles hold another. Dropping the input holder's marbles adds the input number to the toggle number, and you can clear the machine for a new calculation by dropping the result onto the result holder.

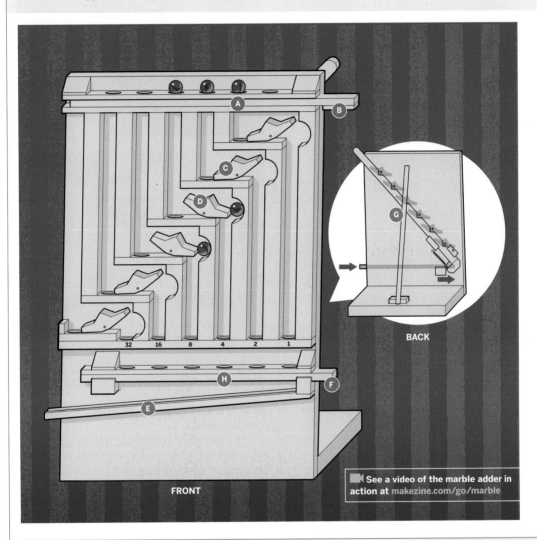

32 16 8 4 2 1

BACK

FRONT

See a video of the marble adder in action at makezine.com/go/marble

The input holder **A** holds the next binary number for the machine to add. The holder's holes are blocked by the input slider **B** until the slider is pushed to the left, allowing the marbles to fall through.

The toggles pivot around nails in the machine's backboard. A marble dropping onto an empty toggle **C**,

representing a 0, rocks it to the right and is held in its cavity, incrementing the corresponding bit.

A marble dropping onto a filled toggle **D**, representing a 1, rocks it to the left. Two things happen as a result: the captive marble is released and falls through to exit down the ramp **E**, and the dropped marble

carries to the left to increment the next bit up. Voilà, binary addition!

The carry marble is delayed compared to the dropped marble, so they don't collide.

When you push the result slider **F** to the left, it pushes a slider on the back of the machine **G** that resets all the toggles to 0

and drops their captive marbles. Meanwhile, it holds the result in the result holder **H** to be read.

Pull the result slider back to the right, and the marbles fall through and clear out, down the ramp. The machine is ready to perform a new calculation.

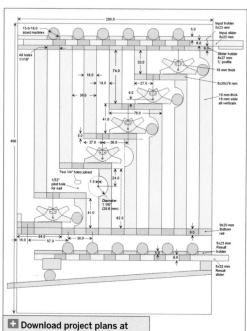

✚ **Download project plans at**
makezine.com/20/marble_adder

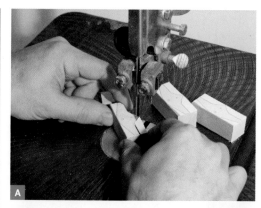

A

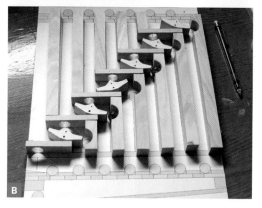

B

MATERIALS

⁷⁄₁₆" plywood at least 400mm×280mm **for the backboard. I used birch.**

Closed-grain hardwood 1×4 (¾"×3½"), 4' long **for the toggles and horizontal pieces. I used birch because I like its light color, but maple would also work.**

Wood 2×6 (1½"×5½"), 4' long **for the vertical pieces and base. I used spruce.**

Carpenter's glue and sandpaper

1¼" finishing nails (13)

Washers (6)

Wood screws: #4×¾" (4) and #6×1½" (4)

Small, loose spring or rubber band

Marbles, 15mm–16mm in diameter (12)

TOOLS

Computer with printer and paper **to print out plans from** makezine.com/20/marble_adder

Band saw or jigsaw

Table saw

Drill or drill press

Drill bits: ¹⁄₃₂", ³⁄₆₄", ³⁄₃₂", ¼", and ¹¹⁄₁₆" **The ¹¹⁄₁₆" bit, for the marble holes, is an unusual size; try a specialty woodworking supplier like Lee Valley Tools (**leevalley.com**), Woodcraft (**woodcraft. com**), or Rockler (**rockler.com**). Substituting ¾" holes won't align the marbles as well; ⅝" holes can work if you carefully select smaller marbles.**

1⅛" Forstner bit **or fine band saw blade or scroll saw**

Clamps

Keyhole saw or file

⟫ Build Your Marble Adder
Time: **A Weekend** Complexity: **Moderate**

1. Mark the layout on the backboard.

You can use my project plans, or you can design your own (see sidebar on page 85). Note that I use metric measurements in general, but standard sizes for the drill bits and some materials.

1a. Download the project plans from makezine. com/20/marble_adder and print them at full size, not scaled. If you can print oversize, use the full template *plan.png*. Otherwise, print out the tiled version, *plan_p1* through *plan_p4*, and align and glue the 4 pieces together. Also print at full size the templates *toggles_template.png*, for Step 2, and *sliders.png*, for Step 3.

1b. Cut the plywood to 400mm×280mm and clamp the template centered on top.

1c. Transfer the key locations from the template onto the plywood. For the positions of the holes, I used an awl to punch through the template and into the wood. For the horizontal and vertical pieces, I transferred the positions of key corners by

Diagrams and photography by Matthias Wandel

C

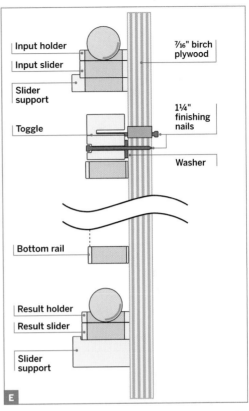

Input holder
Input slider

Slider
support

Toggle

Bottom rail

Result holder
Result slider

Slider
support

⁷⁄₁₆" birch
plywood

1¼"
finishing
nails

Washer

E

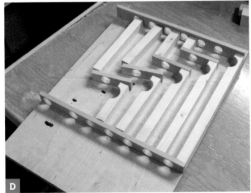

D

punching lightly through the paper with a chisel. Be sparing in how many places you mark, and circle the points on the template where you marked the wood, so that after you remove it, you can still figure out what corresponds to what.

2. Cut the toggles.

Each toggle uses two 1¼" finishing nails: one provides the pivot that it spins around and the other sticks out the back for the reset mechanism. Use a closed-grain hardwood for the toggles and horizontal pieces, so they can withstand all the marble impacts.

2a. Cut out the 6 paper templates from the toggle template you printed in Step 1a, and glue them onto 1×4 hardwood. Use a band saw or jigsaw to cut the shapes out (Figure A).

2b. For the pivot nail holes at the bottom, drill a centered ³⁄₆₄" hole all the way through and countersink it with a ³⁄₃₂" hole 2mm deep for the nail head to sit flush. For the reset nails, drill a ¹⁄₃₂" pilot hole from the back, going only 12mm deep.

2c. Sand the template paper off the toggles.

3. Cut the verticals and horizontals.

3a. Cut the 11 vertical pieces, following the plan. I used spruce 2×6s. All are 19mm thick, front to back, and 18mm wide. Start with the 5 top pieces, which are just simple rails.

3b. For the 6 bottom vertical pieces, which have a cutout next to the toggle, cut 3 pieces just over 36mm wide. Using a 1⅛" Forstner bit, drill holes centered 24mm down from the top, then cut the rails down the middle and trim to length. Don't cut the holes with a spade bit, since that probably won't cut cleanly. If you don't have a 1⅛" Forstner bit, you can cut the rounds out with a fine band saw blade or a scroll saw.

3c. Follow the template *sliders.png* that you printed in Step 1a to cut and drill the horizontal pieces: a feeder above each toggle, the input holder, input slider, input slider holder, bottom rail, result holder, result slider, result slider support blocks, and clear ramp. All holes are ¹¹⁄₁₆". I added some extra length to the clear ramp, not on the template, so the marbles could dump into a bowl more easily.

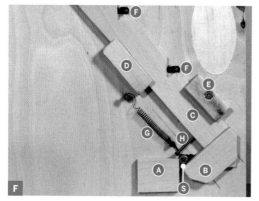

F

G

TIP: As you cut the pieces, check them by laying them down in place on the paper template (Figure B).

3d. Deburr all the marble holes and use a knife or file to make a chamfer on the topside of each (Figure C, previous page). Cutting against the grain will splinter the wood, so each hole requires a separate cut for each quadrant.

4. Assemble the front.

4a. In the backboard, drill the ⅟₃₂" pilot holes for the toggle pivot nails. Above them, drill the holes for the toggle reset nails — two ¼" holes next to each other — then use a small saw or file to join the 2 holes, making an oblong shape. Drill and cut another oblong aperture for the stop screw at the left end of the result slider.

4b. It's time to start gluing pieces to the backboard. First, glue down the input slider holder and bottom rail. Let the glue dry.

4c. Working from one side, glue down sets of 2 uprights and 1 horizontal for each bit position (Figure D).

4d. Glue down the input holder and result slider support blocks (Figure E). To make sure the sliders have enough room to slide comfortably, cut slider-sized spacers out of thin cardboard and sandwich them in with the sliders as you position the new pieces for gluing. Make sure the sliders remain unglued.

4e. Glue down the clear ramp. (So marbles wouldn't fall off my extension to the clear ramp, I added a back piece there, which isn't in the template.)

4f. (Optional) For aesthetics as much as anything, I glued beveled blocks at either end of the input holder and between the bottom rail and result holder. I also added small blocks between each result holder hole. These are not in the template.

4g. Attach the toggles. Nail the pivot nails into the back panels, threaded through the toggles with a small washer underneath. Turn the assembly over and nail the reset nails through the elongated holes into the other side of each toggle so that they protrude about 6mm.

5. Add the release mechanism and base.

5a. Screw a 1½" #6 wood screw into the bottom slider, through its oblong hole in the backboard of the machine.

5b. Make the diagonal slider. It's best to size this to your machine's actual dimensions, rather than from plans. Tilt all the toggles to the 1 position (right in front, left in back). Cut a slat of birch 5mm thick and 23mm wide, and push it up against the nails in back. Use a pencil to mark where the nails touch the slat, then cut vertical notches down and to the right from each, so that they run under the oblong holes.

NOTE: For Steps 5c–5e, the reference position for the diagonal slider is to the left (facing the back), with all toggles in the 1 position (Figure F).

5c. Slide the bottom slider screw to the left in its hole. Referring to Figure F, cut and glue block B to the bottom end of the diagonal slider (C), so that it hooks under the screw.

5d. Mark the left edge positions of the diagonal slider and block B. Cut and glue block A so that it rests against block B as a stop.

5e. Hammer a nail (H) into the bottom edge of the diagonal slider, hook it with a small spring or rubber band (G), and anchor the other end of G to the backboard with a ¾" #4 wood screw, so that it pulls the slider up and left, into the 1 position.

5f. Slide the diagonal slider up and to the right, perpendicular to its length. It should engage the slider screw and toggles so that they all flip to the 0 position. Mark its right edge.

5g. Cut L-shaped blocks D and E to stay the diagonal slider on each side, making sure it will still slide between 0 and 1. Glue D to the backboard and attach E with a ¾" screw, to make the slider removable.

5h. For the base, bevel a piece of 2×6 wood with a 20° angle, then drill and screw the bottom of the backboard to it. I used three 1½" #6 wood screws, spaced out. That's it; you've made it!

NOTE: For reinforcement (optional) you can run a vertical strut from the base to the top of the backboard, as I did with my prototype (Figure G).

Use It! Let's Roll!

Debugging

Chances are, your machine will need a bit of debugging to get it working perfectly. Filing here and there will smooth the marbles' journey.

Hopefully, the 20° angle will keep marbles from getting ejected out the front of the machine, but if they do fall out, it may help to carve out the back edges of the holes that the marbles pass through. You can also just re-bevel the base to mount the machine at a 30° angle instead of 20°.

Adding Value

For maximum marble action, add 63 (111111 in binary) to 63.

It's also fun to use the adder for subtraction and negative numbers by using the *two's complements* of numbers. This means interpreting the leftmost bit as the positive or negative sign, and flipping all the other bits to convert from positive to negative. For our 6-bit adder, this gives a number range from −31 (100000) to 31 (011111). Adding −1 to −1 is the same operation as adding 63 to 63 above, and it results in 111110, the 6-bit two's-complement representation for −2.

In general, it's fun to subtract 1 from numbers, because nearly the same number results, but with all the original marbles dumped and replaced by new marbles.

I invented my marble adder independently, but people have since emailed me about 2 educational games from the 1960s with which it shares some similarities, named Dr. Nim and Digi-Comp II.

Multiplying Possibilities

I have spent time trying to come up with multiplier designs, but everything I've thought of lacks the gee-whiz simplicity of the adder. I'd want to multiply not by adding the number to itself *N* times, but by shifting the number to be added, and adding it to the other argument in each shifted position. Such a machine would no doubt involve a lot of sequencing and require gears and such, but I haven't come up with a specific design that has simple appeal. Suggestions are welcome!

Designing Your Own

If you want to use your own template, try prototyping a 1-bit adder first, since it's tricky getting a toggle shape and configuration that work reliably. To test the whole machine later, put it together using just hot glue for the non-moving pieces so you can pry them off to reshape or reposition them. The MSB (most significant bit) marbles on the left have a long drop, so you should confirm that they stay on track and don't bounce out.

Matthias Wandel (mwandel@sentex.ca) was one of the early engineers on the Blackberry device at Research In Motion. Things have worked out well for him, and now he's an engineer at large.

Getting Back to Nuts and Bolts

ACTOR JOHN RATZENBERGER WANTS TO SEE EVERY KID BECOME A HANDS-ON MAKER.

BY DALE DOUGHERTY

Forty years ago at Woodstock, John Ratzenberger was operating a tractor. He lived nearby in Bearsville, N.Y., working as a journeyman carpenter, and he'd heard the festival was going to need workers. When he showed up, someone asked if he could drive a tractor. "I said yes, and they gave me the keys," he remembers.

What he recalls about Woodstock is the rain and how unprepared the crowd was when the food ran out and the porta-potties were full. "If it wasn't for the National Guard, who arrived with food and toilets, Woodstock would have been remembered as another Donner Party," says Ratzenberger, who stayed busy using the tractor to help free vehicles stuck in the mud.

Ratzenberger is remembered by many as the iconic barfly Cliff Clavin on the TV series *Cheers*, but a younger generation is more likely to recognize his voice. The actor has played toys, vehicles, and other characters animated in every Pixar movie, including Hamm the piggy bank in the *Toy Story* series, and Mack the truck in *Cars*. "I got the part because the folks at Pixar knew my father was a truck driver,"

"I grew up in a factory town and I knew that if you can make, build, or repair something, well, you have to be pretty smart to do it."

MAKER MAKER: (facing and left) John Ratzenberger at the L.A. Community College summer manufacturing camp with more than 20 area teens learning the nuts and bolts of manufacturing. Students at Mid-South Community College in West Memphis, Ark. (above), get hands-on manufacturing experience. Summer camp projects are made with technologies such as welding, CNC lathe, water jet cutter, and lasers.

says Ratzenberger. "That's how they think at Pixar."

"I remember at a very young age being fascinated by the insides of radios," says Ratzenberger, who grew up tinkering in Bridgeport, Conn., where his mother was a factory worker. "My mother would buy old radios at garage sales and I eventually had enough of them and had taken all the parts out of them so then I made a futuristic space city."

Living near the ocean, he and some friends found a boat washed up on the shore. They hauled it home where they recaulked and replanked it. "We were 8 or 9 years old," he recalls. "We didn't think about what we were doing. We didn't call it creative play. We just fixed the boat so we could use it. It was fun."

He wonders if these experiences are still available to kids, and what it means to society if we have fewer people willing to work with their hands. "Every single industry started with one person inventing one thing," he says, challenging anyone to prove him wrong. "Every one of those people started off as a child tinkering. No one wakes up at 32 years old and starts inventing."

He's been talking to members of Congress about a coming "industrial tsunami," which is the title of a documentary he's working on. "If you look at skilled workers in America, from welders to rebar setters and carpenters, the average age is 56 nationwide," Ratzenberger says. "They're going to retire soon

and nobody is coming up after them." He believes America needs more vocational training programs.

The manual arts, as Ratzenberger calls them, go unappreciated in our culture, and he's dismayed that the media often portray skilled workers as "dumb."

"I grew up in a factory town and I knew that if you can make, build, or repair something, well, you have to be pretty smart to do it," he rebutts.

From 2004 to 2008, he made factories the focus of his Travel Channel series *John Ratzenberger's Made In America*, taking TV viewers behind the scenes to see how everyday items are made.

He established the Nuts, Bolts and Thingamajigs Foundation (nutsandboltsfoundation.org) in 2006 to promote tinkering in America and introduce kids to technical trades. Today NBT offers scholarships and organizes summer camps for girls and boys ages 12–16. "Kids can get a first taste at making something themselves," he explains. "Most of them become enamored by their newfound skill."

Ratzenberger says teaching kids hands-on skills is beneficial, even if they don't end up working in the trades. "I am determined to have every parent understand the value of getting their kids making things."

Dale Dougherty is the editor and publisher of MAKE.

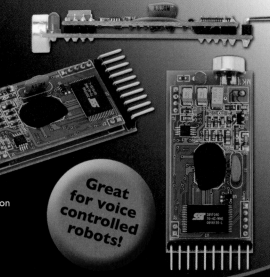

Make: Projects

We're taking classic toys and science projects and giving them a MAKE update. First, fuel a bottle rocket by splitting molecules and going hydro-electric. Next, reanimate a 19th-century parlor toy with a motor and LEDs to give it a high-tech twist. And enjoy 3 laser light shows beamed from your childhood metal lunchbox.

Hydrogen-Oxygen Bottle Rocket

90

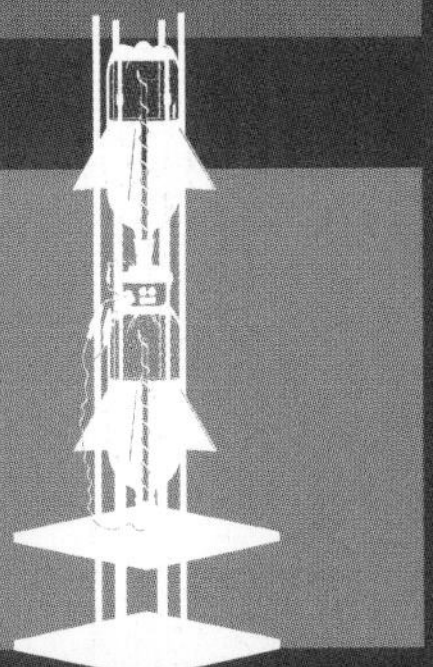

Auto-Phenakistoscope

100

Lunchbox Laser Shows

110

Photograph by Sam Murphy

HYDROGEN-OXYGEN BOTTLE ROCKET
By Tom Zimmerman

FLAMMABLE WHEN WET

Use electricity to split tap water into hydrogen and oxygen gases, then use this explosive gas mixture to power a two-stage, electronically timed rocket.

Air-powered water rockets are easy to build and the fuel is free, but do you want to push the envelope? How about a two-stage water rocket? And instead of pressurized air, how about an explosive gas? I know the perfect gas to use.

Back in high school chemistry, I learned about the electrolysis of water, using electricity to break water molecules into hydrogen and oxygen gases. A spark recombines the elements back into water, releasing energy that vaporizes the water. The hot water vapor causes a rise in pressure, and in this project we use this pressure to push water out of the soda bottles and propel the rocket upward.

This rocket is controlled by an onboard timer circuit that runs off a 9-volt battery and ignites the gases electrically using model rocket igniters. The circuit's variable resistor lets you precisely time the interval between first- and second-stage ignitions to attain maximum height. LED firing signals provide a visual countdown, and when the gases ignite they produce a delightful explosion.

Photograph by Sam Murphy

Set up: p.93   **Make it:** p.94   **Use it:** p.99

Tom Zimmerman is a member of the User Sciences & Experiences Research laboratory at IBM's Almaden Research Center. He graduated from MIT. He was profiled in MAKE, Volume 04.

HYDROGEN-OXYGEN ROCKET — HOW IT WORKS

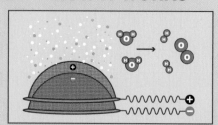

Each water molecule (H_2O) is composed of 2 hydrogen atoms chemically bonded to 1 oxygen atom. Passing electricity through the water breaks these bonds by a process called *electrolysis*. In our gas generator, baking soda makes the water conductive, and stainless steel strainers are used as electrodes that won't rust.

Oxygen atoms are freed at the positive electrode, and they bond with other oxygen atoms to create stable oxygen gas (O_2). Hydrogen gas (H_2) is formed at the negative electrode. Mixing these gases does not form water; at room temperature they remain stable gas molecules. The mixture is known as HHO.

When it's time to launch the rocket, heat from an electric igniter provides the activation energy to break the chemical bonds of nearby gas molecules, freeing hydrogen and oxygen atoms that recombine as the more stable molecule water (H_2O), releasing energy that liberates more atoms that recombine as water, releasing more energy, creating a chain reaction.

The igniter is thin wire with a pyrotechnic chemical on the tip that bursts into flames when current passes through the wire. A dowel positions the igniter above the water and into the explosive gas.

The lower stage (first stage) carries the ignition electronics that ignite both stages in sequence, and guide rails that fit into tubes on the upper stage. When the upper stage fires, its tubes slide off the guide rails and it flies free.

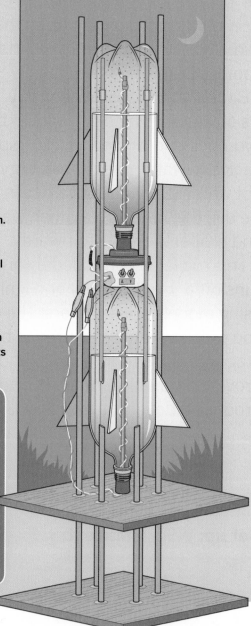

THE IGNITION CIRCUIT

The circuit consists of a 555 timer chip that outputs pulses at about 1 per second to a 4017 counter/decoder. This pulse rate determines the time between the 2 stage firings; the rate is controlled by the 1MΩ variable resistor, indicated by the flashing green LED connected to pin 3 of the 555 output.

The ARM switch powers up the circuit. When the FIRE switch is closed, the 4017 counter/decoder raises the output of one pin at a time, turning on in sequence the green, yellow, and red countdown LEDs to provide a visual countdown (and give you time to run), then it turns on the 2 transistors in sequence to fire the igniters: Q1 to ignite stage 1, followed by Q2 to ignite stage 2.

Illustration by Timmy Kucynda

SET UP.

MATERIALS

FOR THE ROCKET AND LAUNCHER:

[A] 9V battery snaps (2) Jameco #11280 (jameco.com) or RadioShack #270-324 (radioshack.com)

[B] 2-liter soda bottles (2)

[C] Corrugated plastic, 4"×6" (3 pieces) from an art store

[D] Wooden dowels, 36" long: ¼" diameter (3) and ⅜" diameter (6)

[E] Pine boards, 11"×11"×¾" (2)

[F] Insulated solid-core wire, 20 to 24 gauge, 10' RadioShack part #278-1222

[G] Insulated speaker wire, 16 gauge, 20' RadioShack #278-1107

[H] Vinyl tubing, 3" length that slips over ¼" dowel

[I] Sealable, lightweight plastic mini container such as Gladware

[J] Clear silicone sealer

[K] Pencil with eraser

[L] Heavy rubber bands (3) that fit around the soda bottle

[M] Electrical tape

[N] Scrap of foam rubber about 4" square

[O] 9V alkaline batteries (2) one for testing and a fresh one for launch

[P] #4 one-hole rubber stoppers (2)

[Q] Estes Model Rocket Igniters (package of 6) Estes part #302301, from a hobby shop

[R] IC PC board RadioShack part #276-159

[S] Alligator clips (4)

[T] 12V flashlight bulbs (2) RadioShack #272-1127

[U] Electronic components (inside oval):

    8-pin socket, and 14-pin socket

    4017 counter/decoder IC chip Jameco #12749

    N-channel MOSFET transistors (2) Jameco #209058

    555 timer IC chip Jameco #27423

    Mini single-pole switches (2)

    3mm red LEDs (2) Radio Shack #276-026

    3mm green LEDs (2) Radio Shack #276-022

    Resistors: 100Ω (2), 1kΩ (6), 10kΩ (1)

    Variable resistor (potentiometer), 1MΩ

    Capacitors: 0.01μf (1), 1μf (1)

FOR THE GAS GENERATOR:

[V] Oxo brand 8" strainers (2) I found these at Target.

[W] 8" plastic funnel

[X] Heavy-gauge wire, 20' from Jameco or similar vendor

[Y] 5gal bucket

[Z] Large box baking soda

[AA] 12V car battery charger (10 amps minimum) from an auto parts store

[BB] 8-32 ½" bolts (2); 8-32 nuts (2); 8-32 1" washers (6)

TOOLS

⅛" drill

Razor blade knife

Metal ruler

Fine-tooth saw

Hammer and a few nails

Soldering iron, solder

Wire cutter/stripper

Sharpie marker, black

Deep sink or outdoor hose

Protective eyewear

MAKE IT.

BUILD YOUR TWO-STAGE HHO ROCKET

START » Time: **3 Afternoons** Complexity: **Moderate**

1. PREPARE THE GAS GENERATOR

1a. Cut the handles off the strainers.

1b. Using a nail, separate the mesh at the top of each strainer just enough to make room for the bolt.

1c. Place a washer on a bolt and pass it through the hole in the strainer. Place a second washer on the bolt, wrap a heavy-gauge wire around the bolt, place a third washer on top of that, and tighten it down with a nut.

Repeat with the second strainer, mounting the bolts in opposite directions so they won't touch when the strainers are stacked together.

1d. Stack the strainers and glue the rims together with silicone sealer. The Oxo brand strainers have a plastic rim so the metal mesh won't touch when they're stacked; if you use different strainers you'll need to glue foam insulation between the strainers.

1e. Glue the rim of the bottom strainer to the inside bottom of the 5gal bucket with silicone sealer.

1f. Place the funnel on top of the upper strainer. Notice where they touch, and place a few drops of silicone sealer there to glue the funnel to the upper strainer. Let the glue dry overnight.

Photography by Tom Zimmerman

2. BUILD THE LAUNCH PAD

The launch pad can also hold the rocket during construction.

2a. Stack the 2 pinewood boards and clamp them securely, or nail them together temporarily with 2 nails in opposite corners.

2b. Draw a circle whose diameter is ½" larger than the diameter of the plastic bottle, centered on the top board.

2c. Drill six ⅜" holes evenly spaced (60° apart) on the circle.

2d. Unclamp or remove nails, and shove the ⅜" dowels through the holes, flush to the bottom of the bottom board. Slide the top board 6" above the bottom board.

2e. Glue the dowels into the holes with silicone sealer.

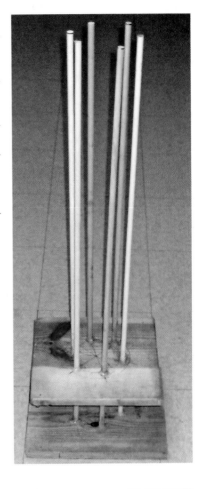

3. BUILD THE ROCKET

3a. Cut 6 fins from the corrugated plastic using the razor blade and metal ruler.

3b. Mark 6 evenly spaced lines (60° apart) on the side of each plastic bottle.

3c. Glue 3 fins evenly spaced (120° apart) on the bottom of each bottle (the end with the nozzle) with the 4" side vertically against the bottle.

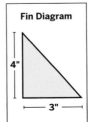

Fin Diagram

4"

3"

3d. Cut six ½" pieces of vinyl tubing and glue them to the bottom and middle of one bottle, 120° apart between the fins. This will be the upper stage.

3e. Glue three ¼" dowels 18" long to the sides of the other bottle, 120° apart between the fins, starting halfway up the bottle. Use the launch pad to hold the rocket during construction. Use 3 rubber bands to hold the dowels in place while the glues dries. This will be the lower stage.

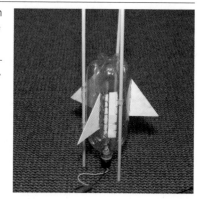

4. MAKE THE IGNITERS

Prepare each igniter carefully because if it doesn't fire, the rocket stage won't ignite.

4a. Cut a wedge into one end of the 10"-long ¼"-diameter dowels. Cut 2 feet of twisted wire, align the wire with the wedged edge of the dowel, and tape the wire 1" below the end of the dowel. Wrap the wire around the dowel, pass the excess through the rubber stopper hole, and then pass the other end of the dowel through the rubber stopper hole.

4b. Strip the ends of the wire taped to the dowel with a razor blade or knife, twist them to the leads of a rocket igniter, and tape them to the wedge with the chemical tip exposed.

5. ASSEMBLE THE STAGE SEQUENCER ELECTRONICS

5a. Use a pencil eraser to clean all the copper on the circuit board.

5b. Solder the sockets in place, then the components. For all electronics connections, refer to the schematic diagram at makezine.com/20/hhorocket.

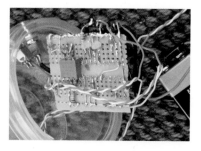

Photography by Tom Zimmerman and Ed Troxell (above this page and bottom facing page)

5c. Put all the LEDs on the same side so you can see them through the plastic case. Put tape around each transistor to make sure they don't touch. Install the 555 and 4017 chips after all soldering is complete.

5d. Check your wiring, then power up the circuit. If the oscillator LED doesn't flash, turn off the power and check your circuit.

5e. For testing, install 2 flashlight bulbs in place of the igniters. Make sure they fire in sequence.

5f. Drill 2 holes in the plastic container for the switches and 1 hole for the igniter wires. Cut the lower stage ignition wires 4" long and solder alligator clips to the end. Cut the upper stage ignition wires 2" long and solder alligator clips to the ends. Seal the igniter wire hole with silicone.

5g. Stuff the electronics, battery, and a piece of foam padding into the case.

5h. Glue the electronics case onto the top of the first stage. This is where you'll clip the electronics to the igniters before launch.

6. GENERATE THE GAS

Two people are required when making the gas; one with dry hands controlling the power supply, the other handling the rocket in the water.

> ⚠ **CAUTION: This step involves a shock hazard. Responsible adult supervision is required.**

6a. Put the gas generator assembly in the bottom of the 5gal bucket, with the wires running outside the bucket.

6b. Fill the gas generator bucket with warm water so the top of the funnel is completely covered (about 5 gallons). Mix in baking soda based on the current capacity of your battery charger; 40 grams for 10 amps, 80g for 20A, 120g for 30A, and 160g for 40A.

6c. Attach the wires to the battery charger and plug in the charger. You should see bubbles coming out of the funnel.

> ⚠ **SAFETY WARNING: Never put your hands in the water when the battery charger is plugged into the wall! Even if the charger is off, you could get electrocuted. The gas mixture you are generating is explosive; operate in a well-ventilated place (outside is ideal) away from any flame or source of ignition. Wear safety goggles, and again, keep the power supply unplugged when anyone's hands are put into the water.**

7. FILL THE ROCKETS

We use a technique called water displacement to fill each bottle with explosive gas.

7a. Fill the bottle with water. With the battery charger unplugged, cover the end of the bottle, submerge it in the water, and place it over the funnel nozzle.

7b. Plug in the charger and fill the bottle ⅓ with gas.

7c. Unplug the charger, put a cap on the bottle while it's still underwater, and remove the bottle from the bucket.

7d. Do the next step outside or over a sink because it's going to get wet. Remove the cap and quickly insert the igniter into the bottle. Don't crush the igniter against the bottle. Let enough water leak out so the bottle is ¼ filled with water, then press in the rubber stopper to stop any leak.

8. OPTIMIZE THE IGNITIONS AND FINE-TUNE THE FUEL MIX

The time between the first and second stage ignitions, determined by the clock rate of the 555 timer, can be optimized. Fire Stage 2 too soon and the full energy of Stage 1 will be wasted; fire it too late and the rocket will be decelerating or, worse, might not be pointing up! Adjust the ignition timing with a twist of the potentiometer. If the rocket starts curving back to Earth before Stage 2 fires, decrease the variable resistance to speed up the sequence rate (the flashing green LED). And if Stage 2 fires too quickly, increase the variable resistance to slow down the sequence rate.

The ratio of gas, water, and air is crucial for attaining the highest flight. The air dilutes the HHO gas to limit the pressure; use too much gas and the bottle will explode. The water provides mass to push against; use too much water, and the rocket will waste energy lifting extra water. In my first attempt I used ⅓ gas, ⅓ air, and ⅓ water. Start with this ratio. Later you can try filling with ½ gas and ⅓ water, which will leave ⅙ for air. If your bottle explodes, use less gas and more air.

You can determine the optimum ratio by strapping a bottle to a free-standing bicycle wheel, as shown here. Ignite the engine with the bottle upright at the 9 o'clock position (temporarily held in place with a stick). Time how long the wheel turns after ignition to get an idea of the power of the engine. More accurate results can be obtained using an electronic force gauge (please post what you learn at makezine.com/20/hhorocket).

FINISH

NOW GO USE IT »

FIRE IN THE HOLE!

LAUNCHING

Place the launch pad on flat, hard ground. Connect a fresh alkaline battery and assemble the rocket on the launch pad. Don't hook up the igniters yet.

Wearing safety glasses, open both the ARM and FIRE switches (i.e., turn them *off*). Make sure that no LED is lit; otherwise you could launch accidentally while trying to hook up the igniters.

Carefully clip the igniter leads to the circuit. Keep the alligator clips separated to avoid a short. The rocket is now armed and dangerous.

To launch, close the ARM switch and make sure the green sequence rate LED is flashing about once per second. Make sure the launch area is secure, yell "FIRE IN THE HOLE!" (my favorite part), close the FIRE switch (you'll see the yellow sequence LED go on), then run away from the pad. The rocket should blast off in seconds!

If after 1 minute the rocket doesn't take off, walk back cautiously and turn off both switches. Drain all the water, turn the rocket upside down to release the hydrogen gas, and figure out what went wrong.

YOUR HHO ROCKET PROGRAM

You can reuse the bottles for future launches and re-glue fins or dowels that break on landing. For softer landings, add a parachute.

To improve aerodynamics and increase flight height, add a nose cone. Also try using wider, taller fins with their shape cut to follow the curve of the bottle at the bottom; our 3"×4" fins are small.

Increase the proportion of gas to make the rocket more potent. At some point the bottle will rip; you can test a bottle on its own to determine this limit.

If you're ambitious you could try 3 stages, or several first stage boosters wired in parallel to 1 transistor to fire simultaneously (you might need 2 batteries in parallel to get enough current).

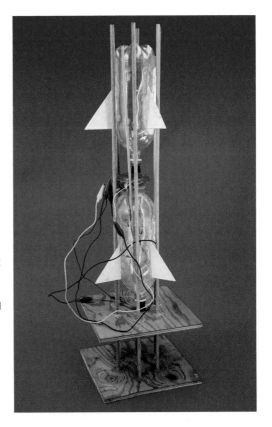

RESOURCES

For the igniter circuit schematic with a list of components and part numbers, visit makezine.com/20/hhorocket.

➕ Here's a good reference on building water rockets: home.people.net.au/~aircommand

🎥 Here's a movie of my first launch: makezine.com/go/hho

Photograph by Sam Murphy

THE AUTO-PHENAKISTOSCOPE

By Dan Rasmussen

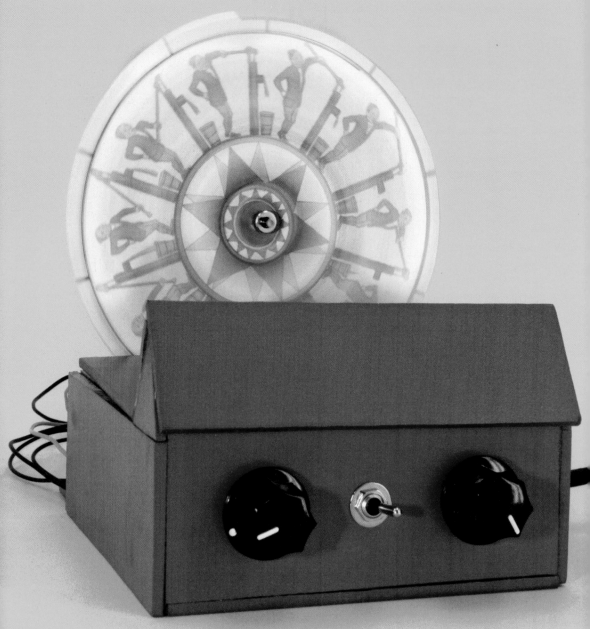

A HIGH-TECH SPIN FOR AN OLD TOY

Motorize a phenakistoscope, a 19th-century parlor novelty that preceded motion pictures, and keep its frames synched to an LED strobe by using a sensor and an Arduino microcontroller.

Invented in 1832 by Joseph Plateau, this device creates a moving picture from a sequence of stills arranged on a spinning disk and viewed through strategically cut slits. It's easy to see how this invention and others like it evolved into movies and television.

My 9-year-old daughter and I wanted to make something new with Lego Mindstorms. We had a strobe, and we thought of combining it with a spinning disk to create animation. But it was difficult attaching the disk to the Lego motor, and we couldn't control the motor and strobe speeds precisely enough to get them to match. The project really came together once we started using a gearmotor to give the disk a smooth, finely controllable rotation speed.

With an adjustable-frequency strobe, we experimented with persistence of vision and sympathetic frequencies, and produced several interesting visual effects. Finally, using an Arduino, a simple C program, and a sensor, we automatically synchronized the strobe with the rotating disk. A switch toggles between manual and microcontroller strobe control.

Set up: p.103   Make it: p.104   Use it: p.109

Dan Rasmussen is a senior IT specialist for IBM and holds a B.S. degree in math from UMass/Amherst and an M.S.C.S. from RPI. He has been working as a software engineer and IT consultant for more than 20 years.

Photograph by Sam Murphy

THE AUTOPHENAKISTOSCOPE

The original phenakistoscope, a 19th-century parlor entertainment, required the viewer to hold its spinning slits up to an eye to see the animation. With single-disk versions, the viewer also had to stand in front of a mirror.

The Autophenakistoscope replaces the slits with an LED strobe, which lets everyone enjoy the view, and also lets you experiment with different disk-strobe timings.

HOW IT WORKS

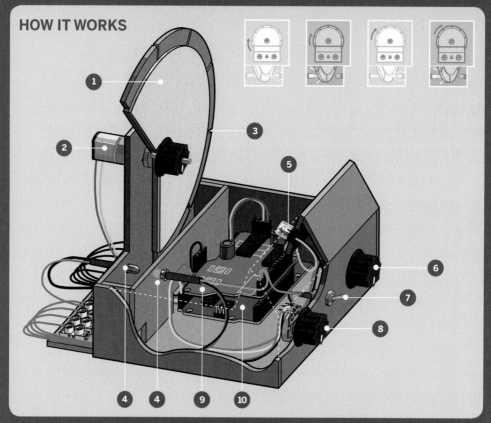

① The image disk has successive frames of an animation loop arranged radially.

② The motor turns the disk.

③ Each slit on the **synchronization wheel**, which surrounds the image disk, sits directly opposite an animation frame.

④ The slits rotate between an **infrared (IR) emitter** behind the sync wheel and an **IR sensor** in front. The slits let IR reach the sensor, which detects when they pass by.

⑤ The strobe flashes to illuminate the image disk. When the strobe is synced to the disk's frames, the animation appears and stays in place.

⑥ The motor knob controls the motor speed in both modes.

⑦ The mode switch toggles between the 2 modes, manual and automatic.

⑧ The strobe knob controls the strobe flash rate in manual mode only. In automatic mode, the sync wheel, sensor, and microcontroller match the strobe to the motor speed.

⑨ The Arduino micro-controller runs a program that receives the sensor input and sends an analog voltage to the motor control board to specify motor speed.

⑩ The motor control board translates control input from the microcontroller into voltages it sends to the motor.

BONUS!

Illustrator Rob Nance created a "Gallopin' CrocoChair" disk to use on your 'scope! Get it at makezine.com/20/autophena

Illustrations by Rob Nance

SET UP.

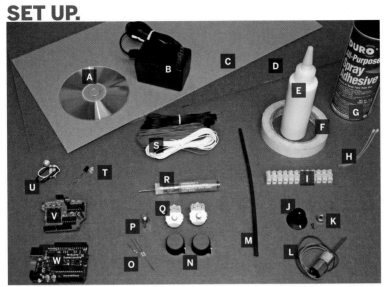

MATERIALS

[A] Unwanted CD or DVD

[B] 12V DC power supply with standard 2.1mm center-positive jack
I used a 1A supply from my junk box.

[C] Non-corrugated cardboard, about 1mm thick, 8"×10" such as mat board

[D] ⅛" hardboard panel (tempered masonite), 12"×16" for the stand and box

[E] Wood glue or white glue

[F] Masking tape

[G] Spray adhesive

[H] Small zip ties

[I] Terminal strip Jameco #215011

[J] Additional knob, or ⅛" shaft collar and washer to secure disks to the motor shaft

[K] 2-56×¼" screws (2) and ⅛" shaft collar for mounting the motor. Use parts #309 and #139 from Du-Bro (dubro.com/hobby), or try your local hardware store.

[L] DC gearmotor I used a 6V DC 488rpm spur gearmotor containing a Mabuchi #FF-030 motor, part #MS-16024-030 from BaneBots (banebots.com), $13

[M] Heat-shrink tubing

[N] Knobs (2) to fit the 5K pots, such as Jameco #265009

[O] Resistors, ¼ watt: 100Ω, 10kΩ (2)

[P] SPST toggle switch Jameco #16523

[Q] 5kΩ potentiometers (2) Jameco #264402

[R] Solder

[S] 22-gauge solid-core insulated hookup wire various colors

[T] Infrared emitter and detector pair RadioShack (radioshack.com) part #276-0142. The ones packaged together in a small plastic bag work better than the ones in a blister pack (both have the same part number).

[U] LED holder/diffuser lens Mouser (mouser.com) part #593-3000GLO and **30,000mcd white LED** This is our strobe; the brighter the better.

[V] Freeduino motor control shield kit for Arduino NKC Electronics (nkcelectronics.com) part #ARD-0015, $11 and **Stackable headers** NKC Electronics #ARD-0021, $2

[W] Arduino Duemilanove Maker Shed (makershed. com) part #MKSP4, $35

[NOT SHOWN]

Snap-in, panel-mount LED holders (2) RadioShack #276-079

Paint

Double-sided foam tape, about 12"

TOOLS

Drill and bits including ¼" brad-point bits

Heavy-duty scissors or compass cutter such as X-Acto #X7753. Electric scissors are optional but handy.

Drafting compass

X-Acto knife

Sandpaper: medium and fine grits

Wire stripper/cutter

Soldering iron

Heat gun or lighter

Pliers

Saw to cut the hardboard clean and square. I used a chop saw, aka miter saw; you could also use a jigsaw with a reverse-tooth blade or a band saw.

Razor saw such as X-Acto #X75300

Hex wrenches

Screwdrivers

Computer with color printer

Sharpie markers, red and black

Multimeter

MAKE IT.

BUILD YOUR AUTO-PHENAKISTOSCOPE

START ⋙   Time: **A Weekend** Complexity: **Easy**

1. PREPARE THE IMAGE DISK AND SYNCHRONIZATION WHEEL

1a. Download and print a sample disk image and the templates from makezine.com/20/autophena. Print the disk image in color, ideally on a laser printer so the colors won't bleed if the disk gets wet.

1b. Trim the disk image with scissors but don't cut out the center hole. Glue it to the CD using spray adhesive. Let it dry for 30 minutes, then cut out the center hole with an X-Acto knife.

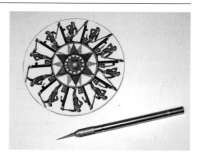

1c. Cut out the sync wheel template but not the slits around its perimeter. Adjust a compass to the outside radius of the template and use it to draw a circle on the mat board. Then cut the circle out of the mat board.

TIP: Cutting mat board with scissors or an X-Acto knife can be difficult, but electric scissors make it easier. The circle doesn't have to be perfect.

1d. Tape the sync wheel template onto the mat board disk and use an X-Acto knife to cut the slits. Remove the template.

1e. Cut the 2 smallest disks from the template out of mat board. The smaller one nests inside the CD's hole and the larger one acts like a washer, reinforcing the image disk in back.

2. BUILD THE STAND AND CASE

2a. Cut the paper stand and case templates, and spray-glue them to the hardboard sheet. Let the glue dry a minute before sticking the paper down, so it's less permanent.

2b. Cut out all the pieces following the templates. I started with a chop saw and finished with a razor saw. To cut the rectangular indentations (wire channels), use the razor saw to cut the sides, then score the top cut with an X-Acto knife and snap it off. Sand all cut edges smooth.

2c. Mark and drill the holes for mounting the motor, IR emitter and sensor, and LED strobe. A brad-point bit will make this easier.

I should have used (but didn't) snap-in, panel-mount LED holders, which mount easily into ¼" holes.

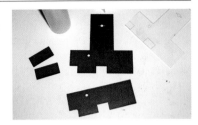

2d. Glue the pieces together to form the stand and case, using masking tape to hold everything together until it dries. Don't glue the front panel cover (that comes later) or the top cover (which is never glued). Be sure the sides of the stand don't protrude from the other segments, or you may have trouble inserting it into the case later. Let dry at least an hour.

2e. Glue the stand into the back of the case. If it doesn't fit, sand down any interference points. Let it dry.

2f. Prepare the hardboard for painting by sanding it all over with medium-grit sandpaper, smoothing imperfections and softening corners. Finish with fine-grit paper, then paint. For bright colors, it helps to start with a primer coat of white acrylic latex. Don't forget to sand and paint the front cover and top cover pieces.

3. PREPARE THE ELECTRONICS COMPONENTS

3a. Cut all the wire segments and strip ⅓" from both ends of each. You'll need:
  12" pieces (14): red (6), black (6), orange, brown
  6" pieces (5): blue, yellow, purple, green, white

The power wires are red and black, and the others are the signal wires. If you don't have enough colors, label the wires as you make the connections.

3b. Solder leads to the two 5K pots, using 12" red and black segments for the outside terminals (for positive and negative power), and 6" unique colors for the signal pins in the middle — I used purple for motor and yellow for strobe. The red and black should run in opposite directions on the 2 pots.

TIP: When soldering wires to posts, first make a mechanical connection by wrapping the wire around the post, then solder the connection.

3c. Solder leads to the IR emitter, which is the dark one in the emitter/detector pair. Solder a black wire to the shorter (−) leg of the emitter, and a 100Ω resistor connected out to a red wire to the longer (+) leg. Seal the connections with heat-shrink tubing by slipping pieces over and shrinking them with a heat gun or lighter.

3d. Solder leads to the IR sensor. Solder a red wire to the long (+) leg and put a 10K resistor between the black wire and the short (−) leg. Also connect a green sensor wire between the negative lead and the resistor. Seal with heat-shrink.

3e. Solder a white wire to the longer leg of the strobe LED and a black wire to the shorter leg.

3f. Connect a red wire to one side of the toggle switch and a black wire to the other through the second 10K resistor. Connect a blue wire between the resistor and the switch. Solder the connections and seal with heat-shrink.

3g. Solder the 12" orange and brown wires to the motor terminals.

3h. Make 2 terminal blocks to organize all circuit connections to positive and negative voltage. Cut a 12-post terminal strip in half, and use markers to color one side red and the other black. For each half, connect a matching lead, and jumper all of the remaining terminals using bare wire. Use a multimeter to confirm that there's near-zero resistance between all terminals on each block.

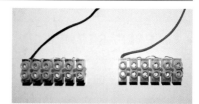

4. ASSEMBLE THE ELECTRONICS

4a. Use two 2-56×¼" or similar small screws to mount the motor to the stand and attach a ⅛" shaft collar to the motor shaft, spaced away from the motor a little so it doesn't hit the mounting screws when it rotates. Verify that the collar is tight and that the motor turns freely.

4b. Mount the IR emitter and sensor facing each other in the stand. I decided to put the emitter in back because it's always on and it connects only to the red/black terminal blocks.

4c. Mount the pots and switch onto the front panel using the included hardware. If necessary, trim the pot shafts to fit the knobs. Install the knobs, and clip any alignment tabs to enable the pots to sit flush against the panel.

4d. Use double-stick foam tape to mount the terminal strips to the floor of the case in back.

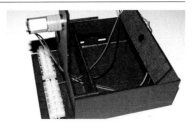

4e. Assemble the Arduino and motor controller boards following the included instructions. On the motor controller, you can use the regular Motors header, but substitute the longer, stackable headers, sold separately, for the Encoder and Power headers. Finally, plug the motor controller board onto the Arduino, making sure to properly line up the pins.

4f. Download the Autophenakistoscope source code from makezine.com/20/autophena, then compile and upload to your Arduino, following the instructions at the same URL.

4g. Make the following connections to the Arduino board:
    Yellow/strobe pot: Analog pin 5
    Purple/motor pot: Analog pin 0
    Green/IR transistor: Digital pin 2
    White/strobe LED: Digital pin 13
        (the LED hangs free, not mounted yet)
    Brown and orange/motor: Shield motor B
        (upper 2 terminals on the Motors header)
    Blue/mode switch: Digital pin 3
    Red/positive terminal block: 5V Power
    Black/negative terminal block: Ground

4h. Run all the red/positive and black/negative leads through the wire channel and connect them to corresponding terminal strips. You can tidy up the wiring with small zip ties.

5. ■ FINAL ASSEMBLY AND TESTING

5a. Cut holes in the center of the sync wheel and the 2 smaller cardboard disks to fit snugly over the motor shaft. Slip the middle-sized disk over the shaft, then the sync wheel (be careful not to bend it), and finally the smallest disk.

5b. Slip the CD onto the motor shaft, facing front. Fit it around the smallest cardboard disk, and secure it so it won't slip. I used another knob to hold it, but you can also use a shaft collar and washer. I drilled the knob to let the long shaft pass through, but you can also cut the shaft.

5c. It's time to run some tests. Connect power to the Arduino, and you should see the strobe flash and the motor spin up to about half power. The strobe should synchronize to the wheel; dim the lights and shine the LED on the image disk to check. Magic!

If it behaves the opposite way (motor goes full speed when turned down low), reverse the Arduino connections for the 2 pots. If it doesn't work, double-check all your connections to the Arduino, and if you still have problems, see my troubleshooting guide at retro-tronics.com/autophenakistoscope.

5d. Once your Autophenakistoscope works, close it up by mounting the white strobe LED to the front panel, gluing on the front panel, and slipping the cover on top. Phenakistastic!

FINISH X

NOW GO USE IT »

USE IT.

TAKE IT FOR A SPIN

TIME AND MOTION STUDIES

Just like the movies, the Autophenakistoscope works best in low light. Dim the lights and plug it in. It first spins up the motor to half power, flashes the LED a few times, then starts reading inputs from the potentiometers. When the motor pot is turned all the way down, the motor should be off. When the strobe pot is all the way down, it should flash about once per second (1Hz).

Adjust the motor and the flash frequency to a moderate level. Toggle the mode switch once or twice to figure out which is which — in automatic mode, the image should stabilize; in manual mode, it will wander or not be legible.

It's time to experiment. Adjust the knobs and you'll see that multiple frequency combinations can produce smooth moving images. It's also possible to freeze the image by adjusting the rotational frequency to match the strobe frequency.

When I first built this project, I noticed that the image can "roll" just like it used to on the old black-and-white TV when I was a kid. It turns out that this is a completely analogous result.

Another interesting experiment is to videotape the Autophenakistoscope. You'll find that the frequencies of the strobe can coincide with the scanning frequency of your recorder. If you get it just right, the scope will appear to be dark in your video when it looks well-lit to your eyes.

DESIGN YOUR OWN IMAGE DISKS

To generate the sample image disks at makezine. com/20/autophena, I took some vintage animation loops and cleaned them up using GIMP software (gimp.org). To make the process easier, I wrote a web application, linked from the same web page, that you can use to make 'scope-ready disks from uploaded photos.

MECHANICAL TELEVISION

I discovered the phenakistoscope while investigating mechanical television. Mechanical television is actually very simple, but impractical except for very small screens. The scan lines are generated with a perforated spinning disk known as a Nipkow disk.

In the 1920s, television started with 24- or 32-line screens, and there were regular mechanical television broadcasts decades before Milton Berle showed his face in our (grandparents') living rooms.

A 420-line spinning disk is another story; it would have to spin at very high frequencies to fully paint the screen at a rate that would satisfy the eye. But the Autophenakistoscope would make a great platform for a modern, experimental mechanical television. I'd also like to see it scaled up.

RESOURCES

➕ For videos of the Autophenakistoscope in action, along with project templates, sample image disks, code, and a web application that generates custom image disks, see makezine.com/ 20/autophena.

➕ Grab a special image disk from project illustrator Rob Nance at makezine.com/20/autophena.

LUNCHBOX
LASER SHOWS
By Mike Gould

COHERENT RAMBLINGS

Back in the 1970s, my friend Wayne Gillis and I used to do light shows at science fiction conventions. We had the usual panoply of overhead, slide, and custom-made projectors, and a single, very expensive, helium-neon laser from Edmund Scientific. Calling ourselves Light Opera, and later, Illuminatus, we performed at ConFusion conventions in Ann Arbor, Mich., and at the World Science Fiction Convention in 1976, where Robert Heinlein was guest of honor.

Flash forward 30 years, and I get a call from David Bloom, a web marketing/connection guy and awesome keyboardist. Would we like to revive our act and perform with him at Penguicon 7? You betcha, I said, and Wayne and I set about to update our craft, revive our spent youths, and order a bunch of now-inexpensive lasers. Thus Illuminatus 2.0 was born.

Penguicon is a unique convention that mixes science fiction fans with those deep into open source software. Given the open nature of the event, we decided to share our tech with MAKE readers interested in laser displays and soldering.

Instead of a single, monolithic laser device, we went with a bunch of inexpensive units, and as "cheapness" was the watchword for this project, we housed the devices in the cheapest metal boxes we could find, namely, lunchboxen. (Penguicon is a Linux convention, so it's one box, two boxen). A search of eBay turned up a raft of cool boxen with science fiction themes, and we were on our way.

Set up: p.113   Make it: p.114   Use it: p.119

Mike Gould is a Mac computer consultant/commercial photographer/technical writer/webmaster/laserist. He and **Wayne Gillis** would like to thank their incredibly patient wives, Salli and Zita, for putting up with all this.

LASER LUNCHBOXEN: HOW THEY WORK

This project includes three laser effects boxen. The Lumia projects a wispy curtain of laser light, the Diffracterator shines multiple swirling beams, and the Motiondizer generates patterns that twitch in time to music.

Lumia

Laser **A** shines into a front surface mirror **B**, which reflects its beam up through the rotating lumia wheel **C**. Patterns or textures in the transparent wheel bend and break up the beam, and because the light is coherent, this generates interference patterns **D** of alternating light and dark. As the laser shines through different parts of the wheel, the wispy patterns change shape. This effect is named after Thomas Wilfred's Lumia performances from the 1930s, in which he reflected colored spotlights off rotating, crumpled metallic foil. Using a front surface mirror (aka first surface mirror) prevents the "ghosting" that results when light reflects off a conventional mirror's front surface as well as its silvered back.

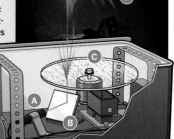

Diffracterator

Laser **A** shines through a rotating, transmissive diffraction grating **B**, which splits it into multiple beams. Each sub-beam then bounces off rotating, reflective diffraction grating **C** which further splits the beams to create a dense, swirling spread. The beam array then shines through the Wobbulator **D**, a transparent, rotating disc with a warp that bends the beams. The result is an enveloping star-field of laser dots that periodically bend and shape-shift, then return to normal motion, possibly inducing vertigo, but in a good way.

Motiondizer

Small front-surface mirrors are glued to a pair of stereo speakers **B** and **C**. As the speakers play music, the mirrors vibrate. Laser **A** shines a beam that bounces from one mirror to the other along a wiggling trajectory determined by the speakers' movements. One speaker **C** is angled so that the beam exits straight up.

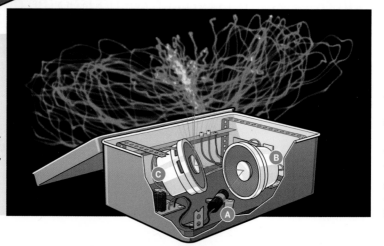

Illustrations by James Provost

SET UP.

MATERIALS

For each lunchbox:

[A] Metal lunchbox about $12 on eBay

[B] ⅜" plywood to fit the bottom of the lunchbox

[C] Metal flat or angle stock for the side rails. Perforated Hyco bar is handy.

[D] 12V DC power supply with connectors wall-wart or bench supply

[E] Small laser module or disassembled laser pointer

[F] Diode, 1N4003

[G] Capacitors: 10µF electrolytic, 0.1µF

[H] Perforated metal pipe hanger strap about 3'

Pipe hangers Sioux Chief ½", ¾"

[I] Phillips-head sheet metal screws

[J] Paint, matte black

[K] Heat sinks and/or 12V DC cooling fans (optional) for more powerful lasers (over 10mW) and their voltage regulators

For each voltage circuit (1 per laser, 1 per motor):

[G] Capacitors: 10µF electrolytic, 0.1µF

[L] Electronic components:

LM317 voltage regulator

Resistor, 240Ω

Potentiometer, 2kΩ We used micro pots, but regular pots are easier to adjust. You can also use a 1kΩ pot for a narrower output range of 1.25V–6V.

[M] Insulated hookup wire

[N] Perf board a small piece

For the Lumia:

[O] DC gearmotor that runs about 1rpm at 3V DC. We used a Mabuchi RF-500T, $5 from Skycraft Parts & Surplus (skycraftsurplus.com).

[P] Transparent disk Use a clear plastic cover disk from a pack of CD-Rs or DVD-Rs, or cut a glass disk 4½" diameter with a ⅛" center hole

[Q] Plastic model cement or clear 5-minute epoxy

[R] Front surface mirror, 1½" square $4 from Edmund Scientific (scientificsonline. com), part #3052323

[S] Washer to fit the motor shaft

For the Diffracterator:

[O] DC gearmotors (3) as above

[Q] Plastic model cement or clear 5-minute epoxy

[T] Transmissive diffraction grating, matrix (aka double axis) $3 a sheet from Rainbow Symphony (rainbowsymphony.com)

Reflective diffraction grating, matrix, 1¼" round $12 for a 57mm x 42mm piece, Dragon Lasers (dragon lasers.com) part #DGRM.

[U] ⅛" plexiglass, 5½" round

[V] Scrap of plastic

[W] White nylon spacer, ½" diameter, ½" long

For the Motiondizer:

[X] Small matched speakers (2) and amplifier Find a used set of powered computer speakers for $5 or so.

[R] Front surface mirrors, 1" round or square (2) Get the lightest weight you can, to improve sensitivity.

[Y] Sheet cork, 2" square We used a drink coaster.

[Z] Thin rubber membrane from a glove or heavy balloon

[Q] Silicone adhesive and 5-minute epoxy

DC power jack (optional) to match the speakers' wall-wart

Panel-mount RCA jacks (2) (optional)

TOOLS

Pliers, needlenose pliers, and tweezers

Tinsnips or hacksaw, file, and hammer for metal work

Pop rivet tool and rivets

Bench vise doubles as an anvil to pound out the curvature of the strap hangers

Saw to cut the plywood base

Drill and bits

8-32 tap and drill set, with tap handle for Diffracterator

Propane torch for Diffracterator

Paintbrush if not using spray paint

Soldering station: Iron, solder, damp sponge

Voltmeter

Wire cutter/stripper

Phillips screwdriver

Pipe cutter (optional) to cut open laser pointers

Toothpick

MAKE IT.

BUILD YOUR LASER SHOW LUNCHBOXEN

START ⟫⟫   Time: **1 Weekend per Box** Complexity: **Moderate**

1. ASSEMBLE THE VOLTAGE REGULATORS

1a. Count the number of voltage regulator circuits you need to build your boxen. Each Lumia box needs 2, a Diffracterator needs 4, and a Motiondizer uses 1.

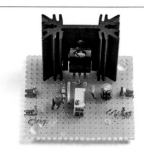

1b. Cut a section of perf board for each box and follow the schematic at makezine.com/20/lunchboxen to assemble its voltage regulator circuits. If you're using a laser stronger than 10mW (or want the option to swap one in later), put a heat sink on the LM317 voltage regulator. Note that all the individual regulator circuits share their connections to power, diode D1, and capacitors C1 and C5.

2. MAKE THE INTERNAL FRAMES

2a. For each box, start with a base plate made of ⅜" plywood. Measure the inside of your box at the bottom and cut a piece of plywood to fit, giving it ⅛" or so of space all around to make removal easier. All the gizmo-ry gets built onto the plate.

2b. Attach side rails to the plate, sized to reach the top corners of the lunchbox. The rails keep everything from falling when the box is on its side, and they also provide places to hang components and serve as convenient handles for removing the workings from the box. I made mine from some erector-set-looking metal stock screwed into ½" blocks of wood glued to the base plate, with the corners secured with pop rivets.

2c. Paint the base plate and rails flat black to prevent random internal reflections.

Now you're ready to make some laser lunchboxen! Instructions for Lumia, Diffracterator, and Motiondizer follow. The Lumia is a good place to start because it's the simplest; all you need is 1 laser, 1 motor, 1 lumia wheel, and 2 voltage regulators. For all boxen, the simplest versions use a smaller laser, say in the 5mW neighborhood, which won't require a heat sink and fan.

3. BUILD YOUR BOXEN

THE LUMIA

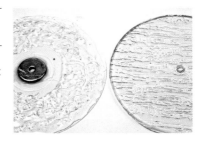

L1. Make the disk, from plastic or glass. For plastic, glue a washer that fits the motor shaft to the center of a CD spacer. For glass, cut the disk (or have one cut) with a glass cutter and glass drill bit.

In a well-ventilated area, apply model glue or clear epoxy to one surface of the disk, smearing it around with a toothpick to make an irregular surface. These disks can be an endless source of experimentation and amusement as you try different combinations of glue, plastic, glass, and whatnot.

L2. Figure out your component layout on the base plate. If you're working with a higher-powered laser, leave space for heat sinks and/or fans.

For the motors and laser, position them to minimize footprint and maximize adjustability, so you can fine-tune the beam paths until you get everything dialed in. I usually screw in just one end of the mounting straps, and only screw in the other end after everything is lined up.

With our Lumia device, we used a big 250mW red diode laser from Sunclan (clantechlaser.com), so the laser and LM317T regulator are both heat-sunk, with a fan blowing on both. The laser is aimed at a front surface mirror, following the usual configuration, but if your laser is small enough you may be able to skip the mirror and just aim it straight up through the wheel.

L3. Attach everything to the base plate, and adjust. To attach the circuit boards, we used 2 methods. I mounted mine directly to the base plate using wood screws; the plywood acts as an insulator. Wayne mounted his vertically using small angle brackets, which minimizes footprint and provides access to the bottom if you need to tweak the wiring. With pipe straps, you can use the existing perforations for screws, and drill extra holes if the spacing is off.

Assemble the device outside the lunchbox, but periodically put it in to make sure everything fits. As you go along, make sure the laser is happy; warm to the touch is OK, but if it's burn-your-finger hot, add a fan. Also check that the motor turns at a nice slow speed; <1rpm is best, but adjust to taste.

L4. Finally, attach the lumia disk to the motor, pop it all into your box, and shine on, you crazy diamond!

THE DIFFRACTERATOR

D1. Lay out the Diffracterator's components on the base plate, just as in Step L2. You need to fit 3 motors. I mounted 2 using pipe hangers with long screws through the clamps. Pipe hangers have a smaller footprint than pipe strap and they can hold lasers or motors, as shown here.

D2. Cut the transmission diffractor disk and screw it onto motor 1 with a nut and washer.

D3. Attach the diffraction mirror to motor 2. To make a mounting I bought a ½" length of ½"-diameter nylon spacer (it's predrilled, so take a motor to the hardware store to choose the right inside diameter), and cut it in half with a hacksaw to make 2 mounts. I tapped the holes to 8-32 so they screw directly onto the motor shaft, and used epoxy to attach the diffraction mirror to one of the nylon mounts. Glue the mirror to the end you didn't cut, to ensure it's at a right angle to the spacer.

D4. Make the Wobbulator, a plexiglass disk thick and warped enough to dramatically change the direction of the laser beams. I used a 5½" disk of ⅛" plexiglass with a hole in the middle, which my glass store cut and drilled for me. Take the disk out to the garage and have at it with a blowtorch, heating it until it's soft enough that you can warp a bit of it into a wavy shape, maybe with some bubbles. Then drip a little airplane glue or clear epoxy onto the warped area.

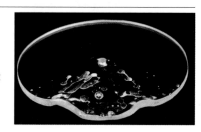

D5. Time for assembly. The Wobbulator takes up the most space, so install it first, screwed onto motor 3, and make sure it clears the side rails.

D6. Remove the Wobbulator and install the reflective diffractor and the laser. Since I used a laser pointer designed for intermittent use, I heat-sunk it by choosing a copper holder and adding copper strip (copper is more thermally conductive than steel). Add the transmissive diffractor. Hook up the power supply briefly to verify the alignment of everything, holding the Wobbulator assembly in place.

D7. Re-attach the Wobbulator assembly. Note that 2 screws hold it down; this is a heavy piece and it needs the support. Finally, add the power supply, hook everything up, plug it in, and admire. The magic begins to happen as the warped part of the Wobbulator does its thing, taking you to the Gamma Quadrant and beyond.

THE MOTIONDIZER

Our box uses a small 5mW red laser. Eventually we'll beef it up — there's lots of room here for more mojo and fans to cool it.

M1. Take apart the powered speakers and remove the amp, which is usually a circuit board in one of the speaker enclosures. Note and label the inputs, speaker outputs, and power input, then unsolder the speakers from the amp.

M2. Prepare the speakers. Bend 2 lengths of pipe strap into L shapes, to use as speaker supports; mine needed 3½" each. Epoxy the back of each speaker to a support, removing paint first if needed. The speaker wire tabs should face up for soldering accessibility, and the base of each L bracket needs room for 2 screws. It also helps to have some extra strap sticking up to use as a handle, and I filed the edges round here, to avoid cutting myself while messing with speaker alignment.

M3. Cover each speaker with a thin rubber membrane, gluing it only around the perimeter. Using silicone adhesive (not model cement) glue a small square of cork to the center of the membrane, and glue a small disk of front-surface mirror to the cork, facing out. If the mirror came covered with protective film, leave it on for now.

M4. Figure out your component layout on the base plate, as in Step L2. Attach speaker C (the one angled up) to the baseplate using just 1 screw (you'll add the second one later once everything is aligned), then bend it back a bit. Similarly attach speaker B (the one the laser will hit) after using a laser pointer to figure out where it should go. With everything lined up, use a pencil to mark the likely locations of the second screws.

M5. Solder your laser to its power supply and attach it where you held your pointer. I mounted it using a pipe clamp that was a bit wide for the laser, so I tormented the flanges until it fit. Verify that the beam travels where it's supposed to and then heads up from the base plate at about a 90° angle. This is an iterative process of tweaking the angles of the laser and both speakers. Once everything lines up, add those second screws to secure the speakers and laser.

M6. Hook up the amp. If your powered speakers run off DC, you can simply use the original stereo input and DC jacks, but to make it neater and more professional we wired these out to 2 panel-mount RCA jacks and a DC jack on the rails. We soldered our jacks to the original jacks' terminals on the underside of the board. Use shielded cable for the input signal between the RCA jacks and the terminals. At the DC jack, split the voltage between the amplifier board and the laser's voltage regulator board. To connect the amp output to the speakers use speaker wire. Observe polarity throughout, +/− and L/R.

If your powered speakers' wall-wart supplies AC, you need to trace the leads on the circuit board through whatever rectifier diodes turn it into DC, and attach your voltage regulator's power wires to the DC rails there. If the original wall-wart supplies 8V DC and you're running on 12V DC, that's no big deal. Underpowering the voltage regulator is OK. Just don't overpower it, as this may result in Bad Things like components melting and explosions.

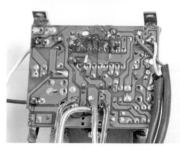

M7. Attach the amplifier and voltage regulator circuit. We screwed the power supply to the base plate, and positioned the amp upright with 2 small angle irons epoxied in place.

M8. If you haven't already, add the rails (for this lunchbox we did this toward the end, contrary to Step 2b). Drill and mount the power and RCA input jacks to an upper rail; it's better to put them here than in the box itself, and it preserves the box's integrity.

M9. Pop the Motiondizer into its lunchbox and test it using any standard stereo output. If it all works, rejoice! Now you can peel any protective film off the mirrors. Note that this setup wants to see line-level signals from whatever is feeding it; see the Use It section for how to run the Motiondizer with separate speakers or a PA for performances.

FINISH

NOW GO USE IT »

USE IT.

SHINE ON

OPEN SOURCE COOLARITY

Wayne and I have made 5 boxen so far: 2 Lumias (red and red-and-green), a Diffracterator (green), and 2 Motiondizers (violet and green).

And as for Penguicon, things worked very well. The DJ at the dance party wanted to buy a Diffracterator on the spot.

I SING THE LUNCHBOX ELECTRIC

Since the boxen run off batteries, you can do shows in camp. Not that you should, of course.

For indoor gigs, Wayne also built a massive power supply into a project case, and I found a suitable snake cable online that we use to get 12V DC to everything that needs it.

We also have several wall-wart power supplies we picked up at ham fests and such. The boxen don't draw much power, so a small wart would probably do you. With a couple of warts, we can do shows without having to drag the big box around.

The Wobbulator disk on the Diffracterator, incidentally, is my one contribution to world culture. There are lots of other laser grating effects out there, but AFAIK, this is the first with a Wobbulator. (Although it may have been invented decades ago by Ivan Dryer at Laserium, I dunno.)

FEEDING THE MOTIONDIZER

The Motiondizer needs full-strength input from your audio system, so if you also want to hear your regular speakers, you may need a splitter such as a distribution amplifier (DA). We use DAs we found on the internet. You can also control the Motiondizer with just a microphone, using a mic pre-amp. In concert, you can run line-level signals for a Motiondizer from the mixing board.

VARIATIONS

Lasers are blisteringly awesome to behold, susceptible to endless tweakage, and dirt cheap. The Lumia and Motiondizer boxes have enough room to mount 2 colored lasers, following slightly different paths to create superimposed pattern projections.

The Lumia projector Wayne built (shown here, minus the wheel) has red and green lasers both cooking away. Note the cooling fan and heat sink on the top laser, a 20mW greenie. The mirror is secured with a cabinet door hinge and a strip of pipe strap.

The motor has a more elegant mounting method: standoffs made for circuit board support. He also used ¼" masonite for his base. Having a better stock of sheet metal than I, Wayne used brass strips and small angle brackets for his laser supports. The blue material around the edge is designed to cushion the apparatus in transit, and the copper strip at the top secures the base plate to the box.

RESOURCES

A good place to start: laserpointerforums.com
For advanced users and professional laserists:
    photonlexicon.com/forums
The pioneering work of Thomas Wilfred:
    lumia-wilfred.org

➕ Visit makezine.com/20/lunchboxen for the circuit schematic, hobby laser and materials buying advice, and photos of the laser lunchboxen in action.

1.+2+3 Remote-Controlled Camera Mount By Ben Wendt

What do you get when you combine a chunky remote control toy car with a lightweight camcorder? You get a street-level action cam that captures video on the move! I came up with this quick and easy mashup for kicks, and have had lots of fun with it. I hope you will, too.

YOU WILL NEED

R/C car with a relatively large base
Lightweight camcorder
Drill with ¼" bit
¼-20 threaded bolt with matching nuts, washers, and wing nut

1. Drill the chassis.
Remove the plastic body from the car and drill a ¼" vertical hole through the chassis.

2. Fasten the bolt.
Push the bolt up through the hole, securing it with nuts and washers above and below. Then put a wing nut at the top, facing upward.

3. Attach your camera.
The ¼-20 bolt fits most camcorders' standard tripod mounts. Screw your camera on, point it in your chosen direction, then secure it by tightening the wing nut.

Use It.
It's really that simple! I've been using the camera to do Hale's Tours of my neighborhood. If you're unfamiliar with the history, Hale's Tours and Scenes of the World debuted in 1905.

Charles Hale used to strap a camera to the front of a train going through a particularly interesting route, and film it. People would then pay to see simulated train rides through exotic locales in this fashion. As an homage to my community and to film history, I've been re-creating this experience using the remote-controlled camera mount.

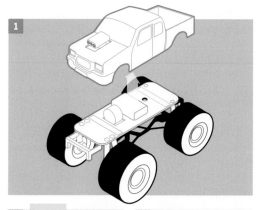

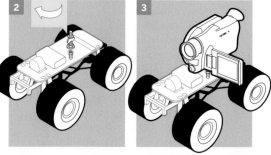

Ben Wendt is a math and computer science dork from Toronto.

Illustrations by Julian Honoré/p4rse.com

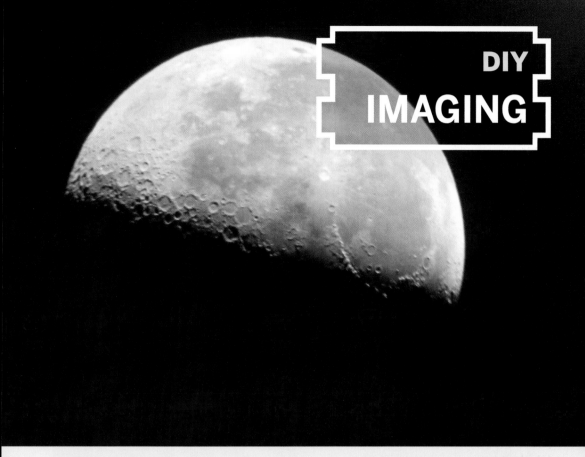

SCOPE PHOTOGRAPHY RIG

Microscope and telescope photography made easy and cheap. By Peter Torrione

Photography by Peter Torrione

Quality cameras for microscope and telescope photography can run $500. But you can also use your family digital camera, attaching it to the eyepiece with a simple mount made from PVC pipe.

1. Prepare the PVC coupling.

PVC pipe couplings are strong, easy to work with, and won't mar the eyepieces. For the telescope mount, I used 1" ID (inside diameter) pipe coupling. Its actual ID measured about 1.3" and it had a ridge in the middle, so I used a lathe to remove the ridge and enlarge the ID to fit around the telescope's standard 1¼" eyepiece. Some patient filing with a half round file will also get you there.

With the microscope, first try holding the camera up to the eyepiece with a tripod to make sure the image can come through. I used a ¾" pipe coupling, which I cemented into a 1" pipe coupling for greater

MATERIALS AND TOOLS

Digital camera with optical zoom It might also need a macro setting. I used a Canon PowerShot SD1100 IS and an old HP Photosmart M425.
Microscope and/or telescope
PVC pipe coupling to fit your eyepiece. A ¾" coupling fits most microscopes, 1" fits most telescopes.
Aluminum flat stock, ¾"×¹⁄₁₆" or 1"×¹⁄₁₆"
#6-32 flathead screws with lock washers and nuts (2)
¼-20 nylon screws (2) or adjustable hose clamp
Spacers or washers as needed
¼-20 thumb screw
Half round file for telescope mount
Drill and drill bits: #34 or ⁷⁄₆₄", #27 or ⁹⁄₆₄", and ⁷⁄₁₆" or bigger
¼"-20 thread tap or hacksaw if you're using nylon screws or a hose clamp, respectively
Calipers (optional) handy for checking diameters and lining up the camera on the mounting bracket

strength, though that's not necessary.

I drilled and tapped holes for two ¼-20 nylon screws to hold the scope's eyepiece in the PVC coupling. This can also be done by hacksawing a slot in the coupling axially (lengthwise), and then tightening it around the eyepiece with an adjustable hose clamp.

Just for looks, I painted my couplings with Krylon plastic bonding spray paint, but it's not necessary.

2. Mount the coupling to the bracket.

The coupling attaches to a piece of aluminum flat stock, ¾"×⅟₁₆" for the Canon and 1"×⅟₁₆" for the HP. Most any hardware store carries these.

Attach the coupling to the flat stock with two #6-32 flathead screws, lock washers, and nuts. Locate these screws at the edges of the coupling, because you'll need to countersink the coupling to accommodate each screw's head and clear the eyepiece.

Put spacers on these screws as needed to center the camera's lens on your scope's eyepiece. For the Canon, I got lucky: a ¼" spacer worked fine. For the HP, a stack of washers did the job.

3. Mount the camera to the bracket.

Bolt the camera to the flat stock with a ¼-20 thumb screw, screwed into the camera's standard ¼-20 mounting thread. Generally, this mount is offset from the axis of the camera's lens, so you'll need to measure accurately to get the assembly to line up.

Also, you must measure the bracket length accurately to accommodate the camera's lens when it's fully extended (not just in zoom mode, when its lens is retracted farther in). This allows the camera to be turned on and off while mounted, with no interference.

4. Shoot through your scope.

Tighten the nylon screws or hose clamp to secure your camera on your scope's eyepiece. Power up the camera and shoot away.

(Of course, since the microscope is fixed, a tripod can also be used to hold the camera — put the microscope near the corner of a table and the tripod as close as possible.)

Peter Torrione is a retired aerospace/defense electronics systems engineer. Since retirement, he doesn't know how he ever had the time to go to work.

A

B

C

D

Fig. A: Telescope mount parts for Canon PowerShot SD1100 IS. Fig. B: Canon mounted to microscope. Fig. C: Canon mounted to telescope. Fig. D: Microscope photo of fly leg taken by HP Photosmart M425 (which also took moon photo, previous page).

INVISIBLE STROBE FLASH

 Near-infrared photography captures bats and other night movers. By Jerry Reed

Photography by Gerald S. Reed

A few years ago, I began documenting the bats flying around my backyard bat house, using inexpensive, monochrome security cameras and a DVD recorder. The little mammals streaked impressively through the frame, but I was disappointed when I started single-framing through to see what they looked like. In each frame showing a bat, it was a mere blur.

I decided to use a xenon strobe to freeze the bats in the images. It wouldn't be synched to the video frame rate, but the combination would be statistically likely to capture at least a few usable stills.

The problem was, some bat experts I consulted thought the strobe would delay their activity and might even drive them away. So I needed to make a subtle strobe, which sounds like an oxymoron.

I knew that xenon tubes produce a wide spectrum, and many tinted plastics block visible light but pass near-infrared (NIR) wavelengths that monochrome

MATERIALS

Xenon flash lamp Party and Halloween stores carry 120V AC versions, and auto stores have smaller 12V DC strobes. You can also assemble a kit from Electronic Goldmine (goldmine-elec.com).

12V DC power source if you use an automotive strobe

"Limo Black" window tint film from an auto store

Monochrome security camera Color CCDs are less sensitive to light, and many color video cameras have built-in IR filters that block the illumination we're working so hard to create.

Lens with a C or CS mount to fit the camera Cheap security camera lenses will work, but good-quality used lenses made for 8mm and 16mm movie cameras are better. The CS mount fits new monochrome cameras, but you can use older C mounts with an inexpensive adapter. Avoid electronic zoom/focus and auto-iris; manual focus and aperture (f-stop) adjustment are best. For nighttime videography, everything is wide open anyway.

cameras are sensitive to. After some experimentation, I found that filtering the flash through 5 layers of "Limo Black" window tint film darkens it to a dim purple spark, but illuminates the field of a sensitive video camera from 10 feet or more.

1. Set up the strobe and filter.

For an automotive strobe, just wrap 4-5 layers of film around the face and duct-tape them together in back. Leave the leads hanging out so you can connect a battery or other 12V DC power source (Figure A).

For plug-in strobes, make a filter by stacking 4–5 squares of film (there's nothing magic about 5 layers; you should experiment). Don't remove the film's transparent backing, which makes it easier to handle. If the strobe has a slot for a colored filter, slide it in there. Failing this, just stick the film over the strobe with duct tape (Figure B). I mounted my camera and strobe to a board to make a single unit. It's ugly, but we're using this in the dark, right?

2. Focus your camera for darkness.

Focusing an IR-sensitive camera for darkness is tricky for two reasons. The longer wavelengths make the focal length longer, due to chromatic aberration. And you're shooting at maximum aperture, which minimizes depth of field.

One way to set the focus for NIR is to focus the camera in daylight with 1–2 layers of tint film over the lens. Another way is to cover an incandescent or halogen floodlight with layers of tint film and use it to illuminate darkened areas. Experiment with the focus and check the video for what works best.

3. Shoot some creatures of the night.

If the strobe rate is adjustable, as with plug-in strobes, set it to the maximum. I'm aiming for bats, but the NIR strobe will work fine for any nighttime subject, moving or not: the neighbor's cat digging up your garden, a passing opossum, and so forth.

4. Post-process the video.

Have patience; the vast majority of frames will be empty, dark, or both. When I find a hit, I bring it into GIMP (gimp.org). First I de-interlace the video, because only one-half (every other scan line) of the frame will be illuminated by the flash. To improve the image, I then usually adjust brightness and contrast and apply an Unsharp Mask (Figure C).

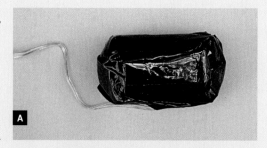

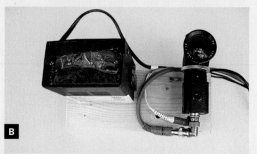

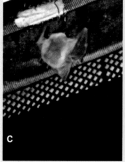

Fig. A: Wrap the 12V DC strobe unit in tinted window film and duct-tape it on. Fig. B: Security camera and filtered AC strobe, mounted on scrap wood. Fig. C: Bat images taken with the NIR strobe.

⚠ Safety Concerns

I'm not keen on "Do Not Use in Bathtub" warnings, but you should not disassemble a strobe unless you know what you're doing, to avoid shocks from its internal capacitors. Also, no one should look at a strobing NIR at close range. Pupils dilate in the dark based on visible light, and the strobe can hit them with possibly damaging levels of infrared.

Jerry Reed is an adjunct professor of computer programming at Valencia Community College, an inveterate electronic tinker, and a friend to bats. karaokebats.blogspot.com

A 1673 VIEW OF THE MICROSCOPIC UNIVERSE

Make your own van Leeuwenhoek microscope. By Patrick Keeling

Antonie van Leeuwenhoek (1632–1723) was a wealthy cloth merchant who lived in the city of Delft, in the Netherlands. He is best known for his pioneering work on microscopy: from 1673 onward he created as many as 500 microscopes, and from these made numerous significant discoveries. It was he who first discovered the existence of single-celled organisms, a discovery that ironically brought his scientific credibility into question for some time.

The success of his microscopes can be attributed to many things, but a number of technical matters stand out. First, his microscopes relied on a single lens. Compound microscopes (those with more than one lens in the light path) theoretically provide better resolution, but they're also much more technically

challenging to fabricate. As well, van Leeuwenhoek devised a method for producing lenses that apparently reduced the need for precise grinding, a laborious and technically difficult process.

The few remaining examples of van Leeuwenhoek's microscopes are elegant creations of brass or silver with many working parts. However, the basic functional aspects of his design and his glass-sphere lenses can be replicated in a few minutes, using simple materials. Following the steps here, you can make a working van Leeuwenhoek microscope capable of 100x to 200x magnification.

It's amazing to consider how we often take microscopy for granted in this day and age. When you use the microscope you've built, imagine what

Photograph by Sam Murphy

Photography by Patrick Keeling

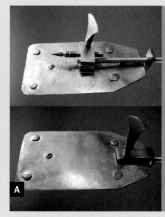

Fig. A: Brass replica of a van Leeuwenhoek microscope (front and back views). You can build one from simpler materials, with the same optics and operating principles.

MATERIALS AND TOOLS

Glass Pasteur pipette or capillary tube or whatever glass source is handy

1mm poster board, 5cm×10cm for the thick side

Cardstock, 5cm×10cm for the thin side

Dab (about 1ml) of tacky poster putty such as Elmer's Tack Adhesive Putty, the kind used to stick posters on a wall. In a pinch, you could use chewing gum.

Drill and bits: ⅟₁₆" and ⁷⁄₃₂" (optional)

Stapler, razor blade, safety goggles

Flame A portable plumbing torch or Bunsen burner works best, but a disposable lighter or even a candle also works if your glass is thin (e.g., a capillary tube).

it must have been like to peer through one of these creations and discover a completely unknown realm of life — because your instrument will reproduce the microbial world just as it would have looked using the technology of the 17th century.

1. Cut the microscope plates.

Cut out 2 roughly equal-sized pieces of poster board and cardstock. They can be any size and shape, but if you want to pay homage to van Leeuwenhoek's microscopes they should be about 6cm×3cm, with a slight taper at one end.

2. Drill the light path.

2a. Drill a 1mm (⅟₁₆") hole in the poster board and cardstock (Figure B). It's best to drill both at once, and to drill into something like wood or additional poster board to give a clean hole. The hole should be 1.5cm from the sides and the top, and about 4.5cm from the bottom.

2b. Make a lens pocket (optional).

If you want to make a little pocket between the poster board and cardstock for your lens, you can drill a shallow depression on the inside face of the poster board using a ⁷⁄₃₂" bit. This isn't necessary, since the cardstock will bend around the lens, but it makes a cleaner finished product.

Alternatively, you can use 2 pieces of poster board, in which case you'll need to drill this pocket or the lens will be too far from the surface to be used (the lens has a short working distance). It's also helpful to carefully shave off any protruding paper around the holes with a sharp razor blade, as loose paper fibers can be magnified along with your specimen.

3. Create the lens.

The lens will be a glass sphere whose diameter dictates the magnification. Aim for a lens about 2mm in diameter since this is big enough to work with and gives a decent magnification.

⚠ **CAUTION: Wear eye protection when melting glass.**

First, stretch the glass. Holding the pipette/tube at both ends, place the center in the flame and hold it there until the glass melts and wobbles freely between your hands. Roll it in the flame to equally expose all sides of the heated area (Figure C). When the glass is soft, remove it from the flame and immediately pull the 2 ends apart to stretch it very thin (Figure D). You're aiming for a glass tube of <0.5mm. Too thick and your lens will be a teardrop rather than a sphere; too thin and you'll have to feed a tedious length of glass into the flame or the lens can break off.

Now form the lens. Once the stretched glass has cooled sufficiently to handle, break it somewhere in the middle. Position the flame horizontally, and slowly feed the stretched glass into the flame from above. Watch carefully. A small, white-hot glass sphere will grow at the tip of the tube as you feed it into the flame. It's critical to keep this sphere in the flame and not to let it cool before it's done — if you pull it out and reintroduce it, bubbles will form. Keep feeding until the sphere is about 2mm in size (Figure E).

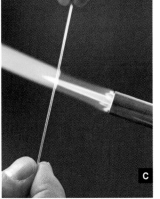

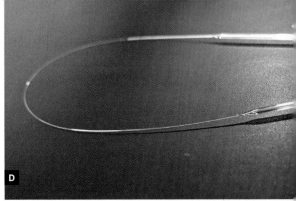

B

C

D

E

F

G

Fig. B: Drill cardstock and poster board at the same time. Figs. C and D: Heat the glass evenly until soft and wobbly, then immediately stretch it into a thin filament or tube. Fig. E: Break the stretched glass in the center, then feed it into a horizontal flame. A small sphere will form. Fig. F: Place the lens in the hole on the inside of the poster board. Fig. G: Staple the 2 layers together with the lens in between.

TIP: A nice trick to help keep this motion constant and the sphere in the flame is to feed the stretched glass though the loop of a pair of scissors so that about 5cm of glass extends from the loop. Then hold the scissors by the blade and twist slightly so there is mild tension on the glass. This will hold the end of the glass steady as you feed it into the flame.

Once the sphere is the size you want, remove it from the flame, let it cool completely, and break the tube off about 0.5cm from the sphere. This gives you a short handle, like a lollipop stick, so you can avoid touching the lens during construction. This also ensures the light path will be perpendicular to the "wound" caused by breaking the lens from the tube.

You can make several lenses from one stretched segment of glass: when one sphere is done, snap it off and start another. They're easy to make, so make several and choose the best sphere with no visible imperfections (such as bubbles).

NOTE: Your lens should be a sphere, not a teardrop. If you get a teardrop shape, your glass was likely not stretched thin enough, so go back to Step 3 and stretch a new one to be thinner.

If you want to work with a smaller lens, you may need to drill holes smaller than ⅟₁₆". The hole must be smaller than the lens, to keep the lens from falling out, and to limit the visibility of your light source background and ensure proper contrast.

If you're going to measure the diameter of your lens for use in calculating its power (see botany.ubc.ca/keeling/resomicr3.html), do it now, before you assemble the microscope.

NOTE: These lenses have very short working distances, so your sample has to be very close to the lens. That's why you use the thinner cardstock on one side. It also means your microscope works in one direction: you look through the poster board side and place your sample on the cardstock side.

4. Assemble the microscope.

Holding your sphere by its handle, place it over the hole you drilled in the poster board (in the pocket if you made one), with the handle lying flat on the inside surface of the poster board (Figure F).

Set the cardstock on top so the lens and handle are sandwiched between the 2 layers. Hold the 2 layers firmly together with the holes and lens lined up,

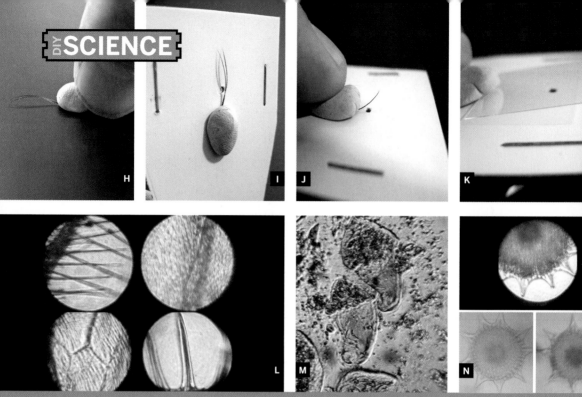

Fig. H: Press tacky putty onto the specimen. Fig. I: Put putty on the microscope. Fig. J: Push or pull putty to pivot your sample closer or farther from the lens. Fig. K: Mount a glass cover slip and drop liquid samples onto it. Fig. L: Insect wing, onionskin, and radiolarian skeleton. Fig. M: Live cells in liquid. Fig. N: Radiolarian skeleton images from a commercial field microscope (left), and 2 homemade microscopes (right and top).

and staple it about 0.5cm on each side of the hole. Staple with the cardstock side up, so the staple is less likely to get in the way of the sample (Figure G).

5. Make your focus mechanism.

The biggest challenge to making a microscope from paper is how to focus your specimen. The solution is to take advantage of the simultaneously elastic and sticky properties of tacky poster putty to both mount the specimen and pivot it in relation to the lens. (Recently used chewing gum works as well if you can't find putty.)

To illustrate how this works, we'll mount the barb of a feather as an example. To start, put your feather barb on a flat surface. Take a dab of putty about the size of a pea, press it over one edge of the barb (Figure H), and pick it up. The barb will stick to the putty and project from it. Now stick the putty below the hole on the cardstock side of the microscope, so the barb is right over the lens (Figure I). Don't get putty on your lens. This will serve as your stage, focusing device, and movement controls.

6. Explore the microscopic world.

To view your specimen, hold the microscope sideways, as close to your eye as is comfortable, looking from the poster board side into a light source, such as a light bulb or a bright sky (don't look directly at the sun). To focus, place your thumb on the putty. Push the putty up toward the top of the microscope to pivot the sample closer to the lens, and pull down to pivot away from the lens (Figure J).

The same principle is used to mount other kinds of samples. Mount any dry, solid specimen as described; to easily see the microscope in action, I like feather barbs, insect wings, or onionskin. To mount a wet specimen, mount a glass cover slip as described (Figure K), then drop your sample onto the cover slip and hold the microscope horizontally with the cover slip facing up (so your sample doesn't drip) and look from the bottom.

Wet samples can also be mounted between 2 cover slips, although this takes a bit of practice. For this I suggest diatoms since they're big and regularly shaped. Forams and radiolaria are also good, but not as easy to find in nature.

Patrick Keeling (pkeeling@interchange.ubc.ca) is a microbiology professor at the University of British Columbia in Vancouver, where he teaches protistology and studies genomics.

BOARDERS AND THE BACKCOUNTRY SUBLIME

 Seeking untouched powder? Mod a snowboard into a splitboard. By Damien Scogin

Snowboarding is a sport of blessed simplicity. One plank. Two edges. Clean lines. But the critical element is the clean lines. From ski-resort brochures to Warren Miller movies, all we ever see are untracked vistas populated by solitary riders who might have simply dropped out of the heavens.

So how can you get to that untracked snow stash? Most ski resorts are surrounded by miles of wilderness, offering the same chutes, bowls, and tree glades, with none of the crowds. This is what we lovingly refer to as "the backcountry." And to get there, you'll want a splitboard — basically a snowboard that splits in two in order to function as a pair of touring skis. No more humping it up the hill with your board on your back while skiers push past you, snickering.

Factory-built splitboards are available, but they're expensive ($600–$1,200 without bindings). If you have an extra board, a prefab kit is available from venerable telemark and splitboard maker Voilé. A number of specialized parts are available only from Voilé, so it's easiest to just purchase the kit. Due to the tremendous abuse that snowboard equipment is subjected to, the connections and hardware must be very durable. Nothing spells trouble like damaged equipment deep in the backcountry.

⚠ **WARNING: The consequences and risks of bodily harm, as well as avalanche danger, only increase as you move away from the patrolled boundaries of ski resorts and rescue personnel.**

Know your limits, get basic avalanche training (AIARE Level 1), and carry avalanche rescue equipment. For more about splitboards and the freaks who ride them, check out splitboard.com.

1. Make the jig and mount the board.

NOTE: If you really trust your skills with a circular saw, then go ahead and use one. However, cutting the board is easily the most critical step in this project, and there's no good way to correct mistakes. For that reason, I chose to use a table saw and build a quick-and-dirty jig to keep the cut straight.

1a. This step is critical. Using a ruler along the base of the snowboard, measure and mark the center in at least 6 places. Using your center marks, take a long straightedge and draw a line the full length of the base sheet (Figure A). Flip the board over and run a strip of masking tape down the center of the top sheet (eyeball it), to reduce splintering.

1b. Cut the 1×10 to the full length of the board plus at least 8" of overhang on both ends. Remove a rectangular section at each end that's at least 11" long and 7½" deep, to accommodate the nose and tail.
   Clamp your 1×10 jig to the board, with its outside edge parallel to the centerline, and its inside edge roughly centered between the factory binding mounts. Mark the jig edge on the top sheet, and mark the locations of 2 factory mounts on the jig. Remove the jig and drill 2 holes at your marks, using a ¼" bit. Be accurate, as you'll be using the outside of the jig against the saw guide. Screw the jig in place using M6×20mm screws (Figure B).

2. Cut the snowboard and finish the inside edges.

2a. Mount the thinnest blade you can on your table saw, to minimize the amount of material removed; I used a 7¼" Skilsaw blade.
   Place your snowboard, base-side down, on the saw table. Support it in place and adjust the blade height so it'll cut through as much of the board as possible, except the metal edges. Adjust the saw's guide to align the blade with your centerline. Repeat at the other end of the board ("measure twice, cut once").

MATERIALS AND TOOLS

Board, kit, and hardware:
All-mountain snowboard with wood core The stiffer the better.
Voilé Split Kit $160 from backcountry.com or voileusa.com
1×10 board at least 16" longer than your snowboard, or ¾" plywood cut to this size for making the table saw jig
West System 655-K G/flex Epoxy Kit 2-part medium-cure epoxy with palettes, gloves, and tools, $28 from westsystem.com
Sanding block with 150-grit sandpaper
M6×12mm T-nuts with teeth (12–16) Call Voilé for these.
M6×12mm flathead screws (12–16) available at good ski and board repair shops
M6×20mm flathead screws (2)

Base repair and tuning:
Swix Base Cleaner for removing excess epoxy and for cleanup
Stanley Surform 7¼" shaver $4 from acehardware.com
Steel base refreshing brush $17 from Tognar Toolworks, tognar.com
Metal Grip P-tex repair string $7 from Tognar
Base repair iron $36 from Tognar
Steel scraper/burnisher $24 from Tognar
Toko board grips (optional) very useful, from Toko (toko.ch) or Tognar
Table saw with 7¼" blade or smaller
Electric drill and bits: ⅛", ¼", 7⁄32", 3⁄16"
Phillips screwdriver
Hex wrench, 6mm
Dremel rotary tool with 2" cutoff wheel
Metal file, center punch, razor blade
Hammer and 13mm socket for T-nut installation
Scotch-Brite scouring pad
Masking tape
Filter mask
Goggles
Gloves

2b. Wear a filter mask, goggles, and gloves to make the cut. With moderate and constant pressure, place the jig against the table saw guide and push the board through. Move the board steadily through the cut and keep any readjustments to a minimum.
   Cut the metal edges with a Dremel and 2" cutoff wheel (Figure C). Any remaining board core must be cut through with a hacksaw.

2c. If you're a stickler, you can mount steel edges into the seam of the cut for better edge performance in ski touring mode, but I chose not to — this is best

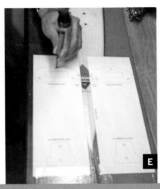
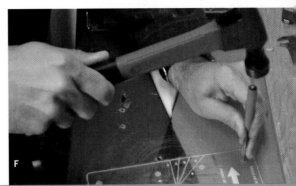

Fig. A: Mark a perfect centerline. Fig. B: This simple jig feeds the board cleanly through the table saw blade. Fig. C: Use a Dremel and cutoff wheel to cut the metal edges. Fig. D: Center-punch or pilot-drill holes for the pivoting board hooks and other hardware. Fig. E: Mark holes for the touring brackets and climbing blocks using the paper template. Fig. F: Use regular or goofy stickers to mount binding pucks at the correct angle.

left to the professionals.

Sand the edges until the seams are parallel and smooth, then clean the edges with a damp cloth and let dry. Mix a batch of epoxy and apply along the edge in a thin layer. Avoid a thick buildup, as it can make connecting the board halves difficult.

Let epoxy cure, then sand and repeat (3–5 coats).

3. Mount the pivoting board hooks.

3a. To maintain alignment over the next few steps, precisely align the snowboard halves, and wrap packing tape around the width of the board in several places, and down the center seam.

With the base on a flat surface, locate and mark the 2 contact points at the nose and tail. From each point, move 1" toward the center (lengthwise) and mark the top sheet.

3b. On the top sheet, position the alignment sticker for the pivoting board hooks (aka "Chinese hooks") at your marks and center it along the seam. Center-punch or drill ⅛" starter holes for accuracy (Figure D), then drill 4 holes with the ³⁄₁₆" bit. Clean excess top sheet material with a razor, and wear gloves to protect yourself from the fiberglass.

3c. Flip the board over and countersink the holes about 3mm deep with a ½" bit, to keep the screw heads flush with your base. Clean up excess base material with a razor.

Remove the sticker from the top sheet and mount the hooks on your screws. The bushings go on the open section of the hook. Tighten the locking nuts down with an 8mm wrench and hex tool — the hooks should swing, but not loosely.

4. Mount the touring bracket.

4a. Place the touring brackets, climbing blocks, and all associated hardware in approximate positions on the top sheet to find an accurate position of the balance point of the board (the extra weight will affect this). Locate the balance point by placing the board on a single Toko bracket or some other fulcrum. Mark this point on the top sheet.

Various sources recommend placing the pivot point ¼" to 1" forward of the actual balance point. This will make the tails drop quicker, thus increasing efficiency in your stride and kick turns.

4b. Align the paper template with your mark and tape it down. Center-punch or drill ⅛" starter holes, 3 for each bracket, front hole first (Figure E). Then drill them with a ¼" bit and remove excess material with a razor. Flip the board over and use a ¾" wood bit to remove the base material — at least 3mm deep. The T-nuts must be sunk into the wood core, and must sit below the surface of the base material. Remove excess with a razor.

4c. Mix more epoxy and apply a healthy glop to each hole in your base. Place M6×12mm T-nuts (with teeth) into the holes and use a heavy-duty clamp to seat them. Repeat for each hole. Place the board on a hard surface, top sheet down, and fully set the T-nuts into the core. This involves a hammer and a sacrificial 13mm socket to do properly. Oh, and a lot of noise. I also recommend a second set of hands to hold the board in place.

NOTE: You may want to drill all your holes at the same time (Figure G), to minimize the episodes of epoxy and loud hammering. If so, skip ahead and perform the placement and drilling stages of Steps 5 and 6. If not, you must let the epoxy cure before doing any further drilling.

4d. Slice off the top part of the template, leaving the climbing block section in place for later (unless you've drilled all the holes at once). Clean the top sheet with the base cleaner and mount the touring bracket using a drill with a large Phillips screwdriver bit. Flip the board over, remove excess epoxy from the connections, and wipe the base clean. Let the epoxy cure. Repeat for the other bracket.

5. Mount the climbing block.

5a. Following the template, drill the marked holes using a 7/32" bit (this is smaller than for the other hardware). Remove excess top sheet with a razor and flip the board over. Countersink with the ¾" wood bit and remove excess material.

5b. Set and epoxy the 10-32 T-nuts as in Step 4c.

5c. The climbing blocks have 3 parts: the shim, the block, and the climbing bar. The shim should be facing the tail of the board; the climbing bar fits into the block and should flip down facing the nose. Clean the top sheet and screw the accompanying hardware into place with 10-32×½" hex screws.

6. Mount the binding pucks and bindings.

6a. Placing the alignment stickers is important, since the blocks provided with the kit don't allow adjustment once they're set. Based on your preferred stance width, find the center of each of the binding positions. Mark the center and note the angle of your bindings.

Remove any packing tape and place the stickers on your marked stance center. Use the appropriate stickers for your stance (goofy vs. regular). Carefully angle the stickers to approximate your binding angles, and avoid wrinkles (Figure F, previous page).

NOTE: Minimum stance width is 18". Make sure the stance angle of your rear binding doesn't come too close to the climbing block (maximum 25°), or it will interfere with getting the slider track off and on. Make sure there's enough room to unclip the pin at the front of the tracks before drilling.

6b. Drill the holes for M6×12mm T-nuts (not the Pozidriv screws supplied with the kit). Make sure you use the appropriate size of wood bit.

NOTE: Here we depart from the kit instructions. Given the terrific forces placed on these binding mounts, and my general paranoia about gear failures in the field, I chose to mount T-nuts through the base just like the touring bracket rather than use the Pozidriv screws provided.
But M6×12mm T-nuts are not widely available. I recommend purchasing them directly from Voilé (they don't offer them online, so you have to phone). There are more widely available T-nuts (M4×6mm or ¼"×½" standard), but these have a shorter collar or a smaller base.

When drilling, you'll probably run into the factory T-nuts for the bindings. If you're lucky, you can use one of these to mount your pucks. But if not, make sure to avoid drilling too close to them and hitting the nut. I simply avoided the hole closest to these and drilled only 3 holes per puck (6 per binding). Using the ¼" bit, drill through the board and clean any excess material. Remove the stickers.

6c. Set and epoxy the T-nuts as in Step 4c (Figure G).

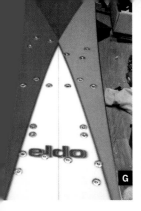

Fig. G: All T-nuts installed for touring hardware and binding pucks. Fig. H: Installing the binding pucks. Note the installed touring hardware. Fig. I: The heel of your binding should be mounted facing the closed portion of the binding plate. Fig. J: Snowboard mode, bindings mounted to pucks. Fig. K: Touring mode, bindings mounted to touring brackets. Fig. L: The climbing bar is very useful for the steeper uphill sections while touring.

6d. The kit comes with 2 sets of pucks (called Nylon Track Location Blocks in the instructions): goofy, and regular. The only way you can tell them apart is the tiny "R" molded in the bottom of the regular pucks. Using a Phillips bit, screw the appropriate pucks into place (Figure H). Slide the metal binding mount on them to ensure that they're parallel across the center seam.

6e. Mounting the bindings is the easy part. The rubber gasket goes between the binding and the aluminum mount (Figure I); if your binding has good dampeners in the base, you can leave this out.

Mount the locking T-nuts through the bottom of the binding mount (the open end is the front) and screw the binding on tight. The screws may need to be cut down to prevent interference with the pucks. Make sure the bindings' toe edges don't extend beyond the front of the mount, as this will affect the touring performance.

Slide the pin through the hole in the front of the binding mount and girth-hitch the cable to the binding. The pin should be easily removable from the mount. For snowboard mode, slide the mount over the pucks (Figure J). For best performance in touring mode, switch the board halves around so the straight edge faces the outside (Figure K).

Remove the binding and binding mount and slide the pin through the holes in the touring bracket (Figures L, M, and N). Make sure the binding buckles face the outside of your "skis."

NOTE: You'll need to buy a set of skins for touring. For steep or icy terrain, there is also a set of crampons that mounts to the binding plate.

7. Mount the tip clips.

7a. Remove the packing tape at the board tips, and place the tip clip alignment stickers. Ignore the rounded dashed line, and place the stickers so the clips will be flush with the tips of the board (nose and tail) and centered on your seam. Center-punch the marks and drill holes with a 3/16" bit, at as close to a 90° angle as you can get. It helps to use a drill block.

Now, flip the board over and countersink the base about 2mm.

7b. Install the rivets, with the head on the top sheet, and the bushing in the slotted side of the clip. Flip the board and, using a thin metal rod (the binding pin is the perfect size, but be careful not to bend it), lightly hammer the rivet to pre-flare it (Figure O).

M

N

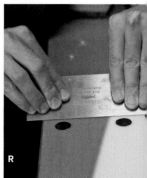

O

P

Q

R

Fig. M: The toe of the binding should face the pin. To switch from touring to downhill mode, simply unlock the pin and slide it from the bracket. Fig. N: Pin locked in touring mode. Fig. O: Set the rivets for the tip clips.

Fig. P: Plug the T-nut holes in the base with hot Metal Grip P-tex repair string. Fig. Q: File down the plugs with the Surform until roughly level. Fig. R: Smooth the plugs perfectly flush with a scraper.

Place a metal block on your work surface and hammer the head of the rivet until the rivet is set into the countersink. If possible, have a partner help you to angle the board so you're striking the rivet at an appropriate angle. Test the connection: the clip should be tight, but able to rotate.

8. Repair the base.

8a. Remove the packing tape holding the board together and pull it apart. It's easier to work on one half at a time. You still have a Swiss-cheese base, so you'll have to do some significant repair work. The key to successful base repair is essentially patience and lots of shaving. Normal polyethylene P-tex repair string and candles do not bind to metal, so you'll have to use a graphite-infused P-tex repair material, like Metal Grip, to bind to your T-nuts.

Cut a 6"–12" piece of repair string and place it close to your hole. Using the base repair iron, heat up the P-tex and glop it into the hole, slowly mixing it in with the flat edge of the iron to remove any air bubbles (Figure P). Build up P-tex to fully fill in the hole, and extend it a bit over the rim. Let it cool to the same temperature as the rest of your base (15–30 minutes). Repeat for the rest of the holes.

8b. Using the Surform, file down the P-tex blobs roughly level with the base (Figure Q), taking care to avoid grabbing any of the excess and yanking out your plugs (the planing heats them up and can weaken the bond). I used a razor to trim back any large bits outside my plugs. Try to plane in the same direction as the base of the board — lengthwise.

8c. Once you've roughly leveled the plugs, use a metal scraper/burnisher to further smooth them (Figure R). Ideally, the plug will feel seamless where it meets the base. Alternate with the scraper, a base brush, and a fine Scotch-Brite scouring pad to smooth the base as much as possible. Although time-consuming, this helps prevent the plugs from ripping out when you remove the skins, and it eliminates drag for better downhill performance.

9. Tune and wax the board.

Numerous online tutorials offer tips and tricks on how to improve your board performance. The best ones I've come across are on tognar.com.

MAKE illustrator Damien Scogin lives in Oakland with his wife and a purple cat. During his downtime, he studies topos for the hidden stashes of California's High Sierra.

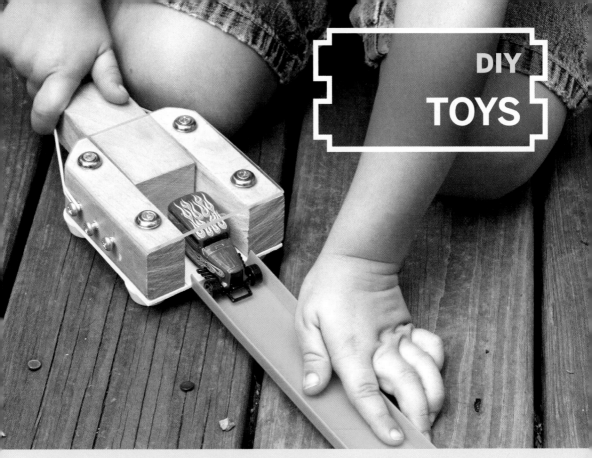

RUBBER BAND POWER!

Build a mini toy car launcher out of scraps. By Nancy "Dot" Dorsner

Photograph by Grieg Wehr

This project was inspired by my son's love of Hot Wheels cars and making them go fast, as well as my husband's love of making cool toys for him and wanting any excuse to play with power tools. This car launcher shoots out Hot Wheels-type cars using rubber band power, and it's makeable using materials you probably have lying around the garage or in the scrap pile.

While I *can* take credit for the concept and documentation of this project, the design and execution were all by my quite handy maker husband, Dameion. This can either be a stand-alone launcher, or can be modified to attach to a standard Hot Wheels track. Here's how to make one yourself, or maybe even two so you can race!

Before you begin, gather the materials and tools listed on the following page.

1. Cut out the launcher.

Find the center of your wood block, center a car on top of it, and mark the car's width just outside each wheel (Figure A, following page).

Use a pencil and straightedge to draw 2 reference lines for cutting, slightly outside the car's width; ours were about 1" in from either side of the block. Cut the block along both lines. Now your wood is in 3 pieces. The 2 side rails will hold the car in place, and the middle block will move to launch the car.

2. Cut and attach the base.

Reassemble your 3 pieces of wood, placing a few business cards between each piece as temporary spacers (cut these in half lengthwise so they don't show and get in your way).

Measure your reassembled block's length and width, and sketch the dimensions onto the plastic for

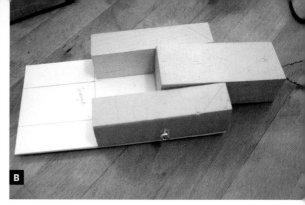

A   B

C   D

Fig. A: Center a car on the wood block and mark the car's width outside each wheel. Fig. B: Use screws to attach the base to the blocks.

Fig. C: Insert the rubber band through the eye screw and stretch its ends over the side screws. Fig. D: Measure and mark your clear plastic to match the top of the block.

MATERIALS

Hot Wheels, Matchbox, or similar toy cars **to ensure size and for testing**
Square block of wood, taller and longer than your cars **Ours was 1¼" tall, 3½" square.**
Short screws (14)
Finishing washers (4) (optional)
Eye screw **aka screw eye**
Scrap of thin plastic as wide as your wood block and a few inches longer **for the base. Or try stiff cardboard, thin wood, or even an old circuit board!**
Scrap of clear, stiff plastic (such as Lexan) as wide and long as your wood block **for the top**
Rubber bands (1 or more) **Various sizes are nice.**
Small rubber or plastic feet **for the base**
String or twine (optional)

TOOLS

Saw **for cutting the wood**
Sandpaper
Pencil
Screwdriver
Drill or Dremel rotary tool
Business cards or scrap cardstock **Try junk mail!**
Utility knife
Straightedge
Sharpie marker **for marking plastic**

the base. For a standalone launcher, mark it 2"–3" longer than your wood block on the front, for a loading ramp. If you'll be attaching the launcher to a track, you can just make the base the same size as the block. Cut the base out, and sand the edges of the plastic.

Drill pilot holes about ½" in from each corner of the base, and centered on each of the side rails. Then drill the holes slightly larger than your screws, so they don't crack the plastic. Use screws to attach the base to the blocks (Figure B). Remove the spacers and make sure that the middle block moves freely.

3. Add the guide screws, and test.

Mark the center of one side, drill a pilot hole, and partially screw in a short screw. Repeat for the other side. These 2 screws will anchor your rubber band. If you like, you can add more screws, to give different levels of tension or to accommodate different-sized bands. We ended up with 3 on each side. On the back of the middle block, mark the very center and screw in a small eye screw.

At this point, it's a good idea to test that everything is working. Insert your rubber band through the eye screw (Figure C), then stretch its ends over the side screws. When you pull back on the eye

Photography by Nancy Dorsner and Grieg Wehr (Fig. H)

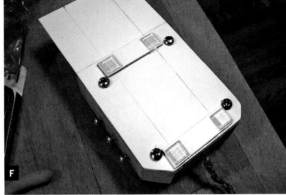

Fig. E: Attach the top as you did the base.
Fig. F: We added glue-on rubber feet to the base to prevent scratching up surfaces.

Fig. G: Pull back the eye screw, put your car in the slot, and let go! Fig. H: The finished track-connector version, with a homemade plastic track connector..

screw, the middle block should slide smoothly in and out.

NOTE: During testing, we decided to add more screws to make it easier for a child to operate. We also trimmed the corners off the back. These steps are optional.

4. Cut and attach the clear plastic top.

Measure and mark your clear plastic to match the top of the block (Figure D). Score deeply along your lines, then snap off the excess plastic. Use a sander or sandpaper to sand any rough edges down to the lines marked.

Attach the top as you did the base in Step 2 (Figure E), making sure to drill the holes larger than the screws. (The washers we used here were for looks only. They're called finishing washers. If you have them, use them, but they're not necessary.)

5. Add the final touches.

» So that it doesn't scratch any surfaces, we added some glue-on rubber "feet" to the base (Figure F).
» A candle rubbed along the sides of the middle

block will make things run smoothly.
» Optional: Decorate with stickers, paint, whatever you like!
» Optional: To attach your launcher to a track, use a track connector as a template to cut your own connector out of plastic. Then super-glue your connector to the bottom of the base (Figure H).

6. Launch some cars.

Attach your rubber band as in Step 3 and pull back the eye screw (we tied some twine to the screw to make this easier for kid-sized hands).

Put your car in the slot (Figure G), and let go! The car shoots across the floor ... Yay!

⚠ **WARNING: Be sure to instruct small children in the care and handling of the launcher before letting them go to the races. Leave it on the ground! No pointing at people!**

Artist and craft blogger Nancy Dorsner can be found at Dabbled: Experiments in Art, Craft, and Food (dabbled.org) when she's not coming up with projects for her maker husband, Dameion.

SAVING YOUR SPECS

Repair plastic eyeglass frames with thread and super glue. By Dmitri Monk

You've got some eyeglasses, plastic frames, broken. On the internet you found a repair place. In two weeks you can get them fixed. But what if you can't wait?

Follow along and fix your glasses at home — the only special tool you need is a commonly available small-gauge drill bit. If you're willing to take a risk and trust in your skills, you can have your glasses fixed as soon as tomorrow!

TIPS ON GLUING:

The secret to good gluing technique is the Three Cs:

» **Cleaning:** Surfaces should be clean and dry with a matte texture. The glue needs a solid, clean surface with a little "tooth" to bond to.

» **Clamping:** Hold the parts steady while the glue sets, or the bond will be weakened.

» **Curing:** Give the glue time to set. If you strain the bond before it's fully set, it will be weaker.

MATERIALS AND TOOLS

Broken eyeglass frames

Nail polish remover or rubbing alcohol

Cotton swabs and a soft cloth

Wax paper (optional) You can use soft cloth instead.

Scraping tool I used a scribe, but hard and pointy is all that matters.

Paint stir stick

Rubber bands Wash and let dry before using.

Super glue Any brand that's cyanoacrylate and runny. Don't use gel.

Very small drill bit 2–3 times the diameter of your needle. I used a #60 wire gauge bit.

Variable-speed drill Slow and gentle is best.

Needle and thread in a color that matches your glasses. Put glue on a test strand to see what it'll look like — silk darkens, polyester doesn't change much.

Fine-grit sandpaper

Hobby knife and masking tape

Photography by Dmitri Monk (this page) and Trudia Monk

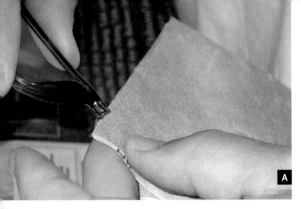

Fig. A: Use sandpaper to rough up the area of the repair and to remove any loose pieces. Fig. B: Make a clamp from a paint stir stick to hold the glasses. Line up the broken pieces and hold them together snugly using the clamp and some rubber bands. Fig. C: Use enough glue so the joint is full, but be neat and don't get glue all over the place. Fig. D: Drill holes on either side of the repair for sewing the tension wrap.

So, to sum up: Clean the joint well, clamp it firmly, and let the glue cure, and you'll have good results.

1. Sand and clean the repair area.

Use sandpaper to de-gloss the area of the repair (Figure A). Lightly sand the break and scrape it with your hard pointy thing, to remove any loose pieces and to make sure it's solid.

Use a swab and nail polish remover or rubbing alcohol to scrub the repair area. Get it clean.

2. Make a clamp.

Cut a piece of stir stick to fit between the temples of your glasses, with some wiggle room. (You'll want some overlap of the broken ends of the bridge, so you can clamp them together firmly.) Wrap it in wax paper or soft cloth to protect your lenses.

3. Clamp the glasses.

Double over a rubber band and slip it over one end of the stick, then slide the end of one half of your glasses under the rubber band. Repeat for the other half.

Scoot the 2 halves together, being careful not to mar the lenses, until the break is lined up and held snugly together (Figure B). Small voids between the 2 sides are OK as long as there are some firm points of contact and things are lined up well.

4. Glue the core joint.

This is the heart of the repair. If you don't make this joint good and solid, the entire repair is at risk.

Use enough glue so the joint is full, but don't let any run out (Figure C). Make sure there are no bubbles, voids, or gaps in the joint. Be neat. Gently roll the side of a cotton swab over the glue to absorb any excess; avoid moving the joint, and blot quickly before it gets tacky.

Gently set the glasses somewhere safe to cure for 1 hour. Keep kids and pets away.

5. Drill holes for the tension band.

Lay the glasses on a soft cloth and drill a hole on each side of the core joint, from top to bottom. Use a hobby knife to dig out little pilot holes before drilling. Place them far enough apart that the repair has a pleasing elongated shape, but not too far apart, or the band might sag.

Drill slowly and carefully (Figure D). Don't stress the core joint or drill holes in your fingers!

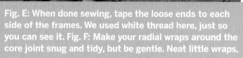

Fig. E: When done sewing, tape the loose ends to each side of the frames. We used white thread here, just so you can see it. Fig. F: Make your radial wraps around the core joint snug and tidy, but be gentle. Neat little wraps, smoothly aligned, are the ideal. Fig. G: Once the reverse radial wrap is done, soak the final wrap in glue. Fig. H: Set the glasses somewhere safe to cure for 24 hours.

6. Sew the break shut.

The tension band is just thread that matches your glasses wound around the core joint through the holes you just drilled. It gives the repair strength.

Thread a fine needle with 4'–6' of thread. Tape the loose end to one side of the frame so that you can wrap snugly. Make your wraps snug and tidy. At the same time, be gentle and don't stress the joint. Sew as many wraps of thread as you can. When the needle won't fit through anymore, you're done — pass the thread across the top and tape it to the opposite side of the frame (Figure E).

7. Glue the tension band.

Use glue to fill the drilled holes and soak the thread, making sure there are no air bubbles, then lightly blot with a swab. Let it cure 15 minutes, and then trim the ends, being careful not to nick the tension band.

8. Wind and glue the first radial wrap.

Tape one end of the thread above a lens, and wrap it carefully and completely across the bridge of the glasses (Figure F). Tape the far end down, but don't trim it. Neat little wraps, all smoothly aligned, are the ideal. A little crisscrossing is unavoid-able, but aim to make the overall effect smooth.

Soak the wrap in glue, making sure it penetrates to the layers below, with no bubbles. Blot lightly and let cure 10 minutes.

9. Wind and glue the reverse radial wrap.

Untape the far end and neatly wind the reverse radial wrap, in the opposite direction. The point is to make the 2 wraps' threads cross over each other to give the repair stiffness and strength.

Tape the loose end down and soak the final wrap in glue, as before (Figure G). Let it cure 2 minutes and then trim the ends neatly.

10. Let it cure!

Set the glasses aside to cure for 24 hours (Figure H). Skimping on cure time will weaken the repair.

Now your glasses are ready to wear. Enjoy!

Dmitri and Trudia Monk are a father-daughter inventor-author-photographer team. Dmitri started disassembling his toys at the age of 5. Now he writes software for a Silicon Valley startup. When Trudia is not attending school or helping fix things, she enjoys training animals.

WIREFINDER 9000

Find buried cable breaks with a radio.
By Christie, Jim, John, and Terry Noe

Photography by Sam Murphy

Our dog, Maggie, loves to run around in our big yard. But how do we fence her in? Building a 500-foot fence is expensive and unsightly, so we use an "invisible fence." This is a buried wire that runs around the edge of our yard. Maggie wears a radio collar that beeps when she gets close to the wire, or gives her a mild shock if she strays over the edge (it doesn't hurt).

One of Maggie's favorite sports is digging up the gophers that are endemic to our neighborhood. Unfortunately for us, the gophers have a hobby of their own, which is biting through the buried wire of our invisible fence.

How do you find a break in a 500-foot buried wire? Digging up the entire 500 feet would not be fun. You can buy commercial cable finders that will find buried wires and wire breaks, like the Armada Technologies Pro 700, but these cost from $500

to $2,000! We decided to make our own and use the money we saved for something useful, like building a big trebuchet.

Overview

Our WireFinder 9000 has 2 components: a transmitter that puts a radio signal on the buried wire, and a radio receiver. You connect the transmitter to one end of the buried wire. This causes a current to flow through the wire, which creates a radio signal that radiates from the wire. Then you can find the wire by walking around with the radio receiver.

When the receiver is close to the buried wire, the radio signal is strong and you can hear it over the receiver's speaker. The strength drops rapidly as you move away from the cable, causing the signal to drop in volume until it becomes inaudible.

If there's a break in your wire, the current can't

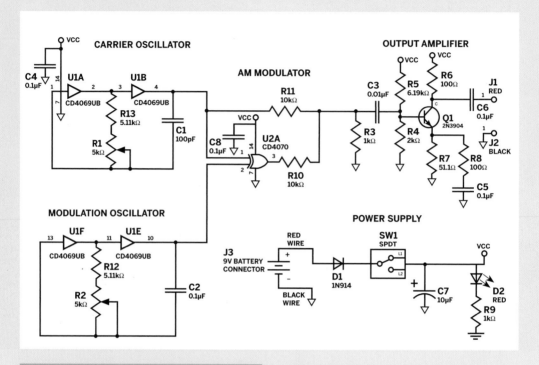

MATERIALS

All parts are available from Mouser Electronics (mouser.com). For a complete list of part numbers, see makezine.com/20/diyhome_wirefinder.

1N914 junction diode **part reference D1**
3mm LED, red **D2**
9V battery connector **J3**
2N3904 NPN transistor **Q1**
5kΩ potentiometers, single turn (2)
   Mouser #652-3386H-1-502LF, **R1–R2**
SPDT push-button switch **SW1**
CD4069 hex inverter, CMOS **Mouser #511-4069U, U1**
CD4070 XOR quad exclusive-OR gate, CMOS
   Mouser #511-4070, U2
Capacitors:
   0.1μF, 10%, 100V (5) **C2, C4–C6, C8**
   100pF, 50V, 5% **C1**
   0.01μF, 50V, 10% **C3**
   10μF electrolytic, 20V, 10% **C7**
Sockets, 14-pin DIP (2)
Banana jacks, panel mount (2) **red J1 and black J2**
Resistors, ¼W, 1%:
   1kΩ (2) **R3, R9**
   10kΩ (2) **R10, R11**
   2kΩ **R4**
   6.19kΩ **R5**
   100Ω (2) **R6, R8**
   51.1Ω **R7**
   5.11kΩ (2) **R12, R13**
Project enclosure and protoboard to fit

reach the isolated segment. So when you move past the break, the signal level drops off rapidly. Using the WireFinder, we were able to locate the break in our wire quite accurately: the place we started digging was only 2 feet from the break.

This technique works for finding both simple wires — like our invisible fence — and buried coaxial cables as well.

To keep things simple and cheap, we chose a regular AM radio as the receiver. Then we designed our transmitter to generate a signal of the right frequency and modulation type so that an AM radio can pick it up. Fortunately, the AM radio band is very low-frequency, which makes construction and wiring very simple. The total cost of the transmitter (Figure A) is about $35.

Construction

Build the transmitter circuit as shown in the schematic diagram (above). Standard wiring and construction techniques can be used. Because AM radio operates at a relatively low frequency, high-frequency construction techniques, such as controlled-impedance lines or coaxial cables, are not required. Figure B shows how we arranged the components on the protoboard.

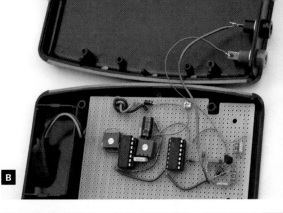

Fig. A: The completed WireFinder transmitter in a slick project enclosure. Fig. B: The completed transmitter, opened up to show component placement. Fig. C: Connect the buried wire or cable directly to the red terminal on the WireFinder transmitter. Fig. D: Follow the buried wire by listening to the tone that it's transmitting to your AM radio. If the tone drops off suddenly, start digging.

Adjustments

There are 2 adjustments you can make in the transmitter circuit. Potentiometer R1 controls the transmit frequency, and potentiometer R2 controls the pitch of the tone you hear on your AM radio.

First, center both pots in the middle of their adjustment range. Tune your AM radio to a vacant frequency at the low end of the tuning range (we chose 574kHz) and place it near the transmitter. Now adjust R1 while listening to the radio. When you tune the transmitter to the same frequency, you'll hear a constant tone in the speakers.

Adjust R2 to change the pitch of the tone to whatever you prefer; we tuned ours to 800Hz.

Using the WireFinder

To find your buried wire, connect one end of the wire directly to the red terminal, J1, on the WireFinder (Figure C). You might find an alligator clip convenient. If you're trying to find a buried coaxial cable, connect J1 to the outer braid or shield of the coax.

Connect the black terminal, J2, to earth ground. The easiest way is to drive an aluminum tent stake into the ground, and connect J2 to the stake. Don't use a painted stake; the paint will prevent you from making an electrical connection. If your soil is very dry, you can improve your ground connection by watering the soil around the stake.

Turn your AM radio on and tune it to the frequency of the transmitter (Figure D). Holding it close to the wire, you should be able to hear the tone. Now you simply follow the wire using the AM radio. If the signal level drops, you've either strayed from the wire path or you've passed the break in the wire.

Eventually, as you follow the wire, you should reach a point where the signal strength drops off in every direction except back the way you came. Now get out your shovel and start digging. The break in the wire should be nearby.

We fixed the break in our wire with special grease-filled wire nuts that are designed for outdoor and underground use in sprinkler systems. You can get these at hardware stores that sell parts for pop-up sprinkler systems. Fix that wire and enjoy the despair of the wire-eating gophers when they realize you cannot be toyed with!

Jim, John, and Christie Noe are high school students in Sebastopol, Calif. They enjoy playing guitar, building trebuchets, and fighting the forces of darkness. They particularly enjoy questioning the sage advice and wisdom that they receive from their father, Terry.

THE CIVILIZED CAT

Train your feline companion to use the toilet. By Josh Klein

Would you let your guests crap in a box on your floor? No? Then why would you let your cats? Here's how to get them using the toilet like civilized members of the family.

Aside from the risks of toxoplasmosis and other diseases transmitted by cat feces, there's significant labor for the owner involved in maintaining a litter box. Being inherently lazy, I decided it would be easier to just toilet-train the cat. There are commercial solutions available (CitiKitty is a good one), but I found it's easy to make an equivalent device from a few items at the dollar store.

One big advantage to doing it yourself is that you can cater the process to your cat's learning curve, which can vary widely from animal to animal.

Functional Overview

The system is a toilet lid with a flat-bottomed bowl attached to its underside. It starts out just like a regular cat box, except nestled in your toilet. Over time you can cut away more and more of the bowl (where your cat will be standing), thus forcing the cat to adapt to standing on the lid and squatting over the hole that contained the bowl. During the training process a dustpan or another bowl can be used to catch the litter when you need to use the toilet.

In designing this system I tried to make it as easy for the cat to use it as possible. One issue many cats have is that squatting on the toilet seat (not the lid) is a slippery, sloped, and narrow proposition. If your cat is at all standoffish about this (mine sure was), then this system ought to work better. As an added advantage, this system allows you to put the lid down so you don't sit on whatever your cat has tracked in on their paws. The components should cost you roughly $10.

Photography by Hulda Emilsdottir

Fig. A: Mark a plastic bowl for cutting into a litter holder. Fig. B: Fit the bowl against the back of the toilet seat lid. Fig. C: Mark bowl position on the lid. Fig. D: Cut a circle into the lid over the bowl location, small enough to let the cat squat in front. Fig. E: Litter holder bowl screwed securely onto toilet seat lid. Fig. F: Sand the front of the lid so it takes glue better. Gluing sandpaper back-side down turns this into a sure-grip platform for your cat.

MATERIALS AND TOOLS

Toilet seat with lid about $3. Cheap lids may be better as they're likely to be thinner plastic, and hence easier to cut.

Flat-bottomed plastic bowl that can be cut without cracking. Get one with as short a slope as possible around its edges — the cat will stand in the bottom. The bowl must be big enough to fit inside the toilet seat and have a lip that extends past the hole in the seat, so that it suspends inside the seat. Get as close a fit as possible — the bigger the bowl, the happier the cat.

Machine screws and nuts (3)

Large-grit sandpaper

Household glue

Sharpie

Dremel rotary tool

Flushable cat litter World's Best works well.

Cat

Construction

This project is easy; overall it took me about an hour to set up.

1. Cut off one edge of the bowl in a straight line across the point where the lip meets the flat bottom (Figure A). Use a Dremel to smooth the sharp edges.

2. Put the bowl into the toilet seat and close the lid. Flip the toilet seat/lid upside down (Figure B) and lift the seat off. Mark where the bowl sits on the lid using a Sharpie (Figure C).

3. Now you'll cut a hole in the lid large enough that the cat can sit inside it and use it as a litter box, but leaving enough space on the front of the lid that the cat can squat there when it's fully trained. Within the marks you made, trace a circle that you think fits the bill — I used a cooking bowl as a guide. Set the circle toward the hinge at the rear to maximize the standing space at the front of the lid.

4. Using the Dremel, cut the circle out of the lid (Figure D). Again, smooth the sharp edges.

5. Drill 3 holes through the lid and the lip of the bowl for the screws— one on each side, and one on the front (non-hinged) side of the lid. The bowl should cover the hole you cut in the lid, with its sheared-off edge facing the hinge.

6. Screw the bowl to the lid (Figure E). Seat it as snugly as possible; if it wiggles or shakes, the cat may feel insecure and not take to it as quickly. Depending on the placement of the screws, you may have to Dremel off the rest of the screw so the lid sits properly on the seat.

7. Use a corner of the sandpaper to sand down half of the topside of the lid, at the front end. This provides some texture for glue to adhere to (Figure F).

8. Glue the back of the sandpaper to the part of the lid you've textured, and let it dry. The flatter the bond, the less likely your cat will find something to object to.
9. Affix the seat and lid to your toilet, put some flushable cat litter in the bowl, and start training!

Operation

You should have your cat using this system within a month (but please see the warnings, at right).

Like most things involving animals, the idea here is simple in principle and difficult in practice; some cats will take quickly to some parts of the process and others may not. If your cat has trouble, go back a step or two until it seems comfortable.

It can be helpful to use treats to reward the cat for jumping up onto the toilet lid. As your cat gets better at this, you can delay giving it the treat until it stands on the correct part of the lid, and once it masters that, until it sits down there. This is called operant conditioning (Google "clicker training" for the latest trends in this process). Here are the basic steps:

1. Move the cat's litter box next to the toilet. If your cat is freaked out by this, move the box back to within 1' of its original location. Move it 1' closer to the bathroom every day until the box is directly up against the toilet.
2. Gradually raise the litter box by placing it on something solid and stable (like phone books); you can tape it in place to secure it. Raise it every day until it's as high as the toilet. If your cat seems comfortable jumping up into the box, speed this part of the process up. If it has trouble, go slower.
3. Move the box over onto the toilet seat and tape it down. Reduce the amount of litter in the box until it's only 1" deep.
4. Replace the litter box with the apparatus you built. Put a 1" layer of litter in the bowl.
5. Every third day or so — or whenever the cat has used the system successfully for 2 full days — Dremel off about 1" from the bowl. Cut it back from the straight cut you originally made, so that the standing surface inside the bowl gradually diminishes toward the front of the lid. This should force the cat to stand more and more on the lid and less and less in the bowl.
6. Once you've cut away all of the bowl, remove it entirely. Voilà — you have trained your cat to use the toilet. You can scatter some litter on the water to encourage your cat through this final stage.

Again, each cat will take to these stages differently. My cat skipped straight to stage 4 and then got very finicky about 5 and 6. Other cats have taken forever to recognize that the litter box had moved to another room, and then immediately started doing their business in the toilet. YMMV, so go as slow as your cat wants to — remember that cat pee smells awful and an annoyed cat can use it on your furniture at any time.

A Word of Warning

There are a few big issues to keep in mind when deciding whether to try to toilet-train your cat:

» **Don't teach your cat to flush.** Once it learns, it will find this an entertaining way to pass the time and your water bill will skyrocket.
» **Don't toilet-train if you live near a body of water where river or sea otters live.** Cat feces can contain a protozoan called *Toxoplasma gondii* that is known to kill otters.
» **This is the big one: If your cat isn't taking to the toilet within a month's time, quit trying** (at least for a few months). Domestic cats are very prone to urinary tract infections, which can kill them if left untreated. Most often these occur when a cat is holding its urine, which it's likely to do if stressed or uncomfortable with its toilet situation. Your cat may simply not be a good candidate for using the toilet, in which case it's better to let the cat have its way.

Signs of urinary tract infection include (but are not limited to):
• urinating in inappropriate places
• crying in the litter box
• frequent urination in tiny amounts
• straining to urinate
• bloody urine
• excessive genital licking
If you observe any of these signs, see your vet immediately.

➕ Visit makezine.com/20/diyhome_cattoilet for additional tips.

Josh Klein (www.josh.is) does systems hacking, including social networks, computer networks, institutions, consumer hardware, animal behavior, and, most recently, the publishing industry. He speaks, writes, and consults on new and emerging technologies that improve people's lives.

ECONOWAVE SPEAKERS

 Turn good vintage speakers into great modern ones. By Ross Hershberger

Last year on the Audiokarma (audiokarma.org) discussion boards, members Zilch and Jackgiff shared a project that rocked the online audiophile community. They designed a treble waveguide and crossover system that greatly improves the sound from older speakers with "fried egg" style tweeters.

Waveguides are horns that disperse high-frequency sounds evenly over a wide area, rather than letting them fall off at the sides. This gives the speakers "constant directivity," which means they sound more natural to listeners who aren't in the center sweet spot, between and in front of the speakers.

Zilch and Jackgiff also designed a replacement crossover circuit that balances the sound across the new combination's optimal 1,600Hz crossover point, routing low frequencies to the speaker's great original woofers and high frequencies to the new

compression drivers. Zilch sells PCBs for building this crossover.

I tried the conversion myself on some 1970s-era Advent Large speakers (Figure A, next page). It cost a couple hundred dollars, but the results were outstanding: the modified old Advents performed like thrillingly clear, state-of-the-art speakers that audiophiles pay thousands for. Since then, I've built and sold a second pair, and many other Audiokarma members have built multiple pairs of these speakers.

1. Take them apart.

Pull off the speaker grilles and then wrench off the grille attachment blocks that are stapled and glued to the baffle. Unscrew and pull out the woofer and tweeter, cutting their wires. Pry off the plastic trim. Wearing a dust mask and gloves, pull out the fiber-glass batting and set it aside in a closed bag.

MATERIALS AND TOOLS

Advent Loudspeakers (2) or similar Any in Advent's Large series will work: Large Advent, Large Advent Utility, or New Large Advent. With the vinyl-covered Large Advent Utility you don't need to trim the sides of the waveguide, but the other versions, with walnut veneer, look cooler. Or you can use any other 2-way speakers with room for 6.4"×11.9" horns and woofers that play up to 1,600Hz.

Pyle Pro PH612 1" screw-on constant radiation horns, aka waveguides (2) part #292-2572 from **Parts Express** (parts-express.com), $13

Selenium D220Ti-8 1" titanium horn drivers, aka compression drivers, 8-ohm, 1⅜"-18 TPI screw mount (2) Parts Express #264-270, $41

ZilchLab EconoWave crossover PCBs (2) from **ZilchLab** (i_am_zilch@att.net), bare board for $20 or fully assembled for additional cost. Or you can wire the crossover circuits on plain breadboard.

L-pad volume controls, 50W mono, 1" shafts, 8-ohm (2) Parts Express #260-255, $10

Resistors: Dayton Audio precision Audio Grade, 10W: 16Ω (2), 30Ω (2) Parts Express #DNR-16 and #DNR-30

Capacitors: Dayton Audio metallized polypropylene, 250V: 0.47µF (2), 4.7µF (2), 12µF (2) Parts Express #027-406, #027-422, and #027-430

Inductor coils: 0.60mH, 20 gauge, air core (2); 1.5mH, 18 gauge, I-core (2) Parts Express #255-040 and #266-552

2-conductor speaker wire, 18 gauge or larger, 9'

Plywood pieces, 5"×5"×10mm thick (2)

Stiff plastic tubing, ¼" or larger, 2" length for screw spacers. You can use the barrel of a stick ball-point pen.

Binding posts, aka speaker terminals, Dayton Audio BPA-38G HD (2 pair) Parts Express #091-1245

Speaker gasketing tape (foam), 1 roll, ⅛"×⅜"×50' Parts Express #260-540

#6×⅝" sheet metal screws with matching washers (20)

#6 wood screws: ¾" (4), ⅝" (8)

Refoaming kit (optional) if the foam is decayed around your original woofers. I recommend the kits from Rick Cobb, rcobb@tampabay.rr.com.

Screwdrivers: slot and Phillips

Wire cutter

Utility knife

Needlenose pliers

Coarse sandpaper around 100-grit

Broad chisel, scraper, or putty knife

Soldering iron and solder

Adjustable wrench

Drill and bits for wood

Small hand jigsaw or coping saw

Painter's tape

Sharpie marker

Large plastic bag

Dust mask and gloves

Band-Aids

A

B

Pull out the staples and chisel off the glue inside to remove the original crossover circuit (Figure B). Sand off any glue residue. Repair and refinish the cabinet as desired.

TIP: Never use steel wool on a loudspeaker; it sheds steel particles that clog the driver's magnet gap.

On old speakers, the foam around the woofers may have cracked and disintegrated (Figure C). This compromises the air seal essential to the bass performance and makes the woofer rattle and sound bad. Fortunately, it's 100% repairable. You can use a kit to replace the foam surround yourself (see Materials), or take them to a repair shop; I like Audio Atlanta (audioatlanta.com).

2. Cut the waveguide holes.

In each speaker baffle, cut a 10½"×5" clearance hole for the new waveguide, aligning the top edge with the existing tweeter hole. I taped off the area, drew cut lines, and used a jigsaw with a 3" blade (Figure D). Tape over the cabinet rim and hold the jigsaw above it to avoid scratching it. Drop the waveguide in. Mark and drill the ten ¹⁄₁₆" mounting holes.

Fig. A: Large Advent, Large Advent Utility, and New Large Advent. Fig. B: Stripped cabinet. Fig. C: The woofer surround's decayed foam is easily replaced. Fig. D: The speaker baffle, cut out for the waveguide and drilled for screw holes. Fig. E: Assembled cross-over boards, with L-pad wires in yellow/red/black and speaker wires in gray. Fig. F: The crossover board and plywood terminal board, mounted in the cabinet.

3. Build the crossover circuit.

Assemble a crossover circuit for each speaker, referring to the schematic at makezine.com/20/diyhome_econowave. It's easiest to build them on the printed circuit boards from ZilchLab, which come with instructions, but the circuit is simple enough to wire together neatly on plain breadboard.

Solder three 8" single wires running offboard for the L-pad tweeter volume control. The L-pad uses 2 potentiometers, a series and a shunt, to let you adjust the tweeter's volume without changing its overall impedance. Also solder two 24" 2-conductor speaker wires for the woofer and tweeter, and mark the positive (+) lead on both (Figure E).

Crimp the slide-on connectors provided with the compression driver onto the tweeter leads.

4. Mount the electronics.

The crossover circuit, L-pad, and speaker terminals are mounted to a 5"×5" square of plywood inside the cabinet. Drill a ⅜" hole through the plywood for the L-pad knob shaft, and 9/32" holes spaced ¾" apart for the speaker posts, close enough to the L-pad so they'll all fit through the cabinet's original terminal hole in back. Also drill a ⅛" hole at each corner of the plywood panel, for mounting it.

Solder the L-pad wires to the L-pad terminals, following the schematic and labels 1, 2, and 3 on the board. Rim the cabinet's terminal hole with foam gasketing tape, and then mount the plywood panel to the speaker with screws. Position the crossover board on the inside of the panel, and screw on the knob and posts on the other side, using washers if necessary (Figure H, next page). Secure the crossover board over the speaker terminal shafts using the supplied nuts. Finally, screw down the free edge of the crossover board, opposite the terminal shafts, with #6×¾" wood screws and plastic-tubing spacers cut to size (Figures F and G).

5. Final assembly.

Pull the speaker leads through their baffle holes, leaving plenty of slack inside, and tape them to the edge of the cabinet. Carefully replace the fiberglass batting, loading it evenly. Replace the cheesecloth behind the woofer hole, and rim the woofer and waveguide holes with more foam gasketing tape.

Screw the compression driver to the waveguide, connect the tweeter leads, and mount the waveguide to the baffle with ten #6×⅝" sheet metal screws and #6 washers.

The waveguide fits the Utility model without

Fig. G: The side view of the terminal board. Fig. H: The finished terminal board, as seen from the outside. Fig. I: The plastic edge on the New Large Advent must be trimmed slightly to clear the waveguide.

Fig. J: Finished speaker. Fig. K: Finished speaker with new grille. Fig. L: An alternate version, with back-mounted woofer and a fancy finish, including a new veneered baffle board with brass trim.

trimming, but the other speaker models require some more work. With the Original Large Advent, sand ¹⁄₁₆" (or less) from both sides of the waveguide to make it fit. The New Large Advent has a bit less space, so you can trim ¼" off each edge of the waveguide, or get tricky with a router and undercut the walnut rim to make room to slide the waveguide edge underneath. I've seen both done, and I'd just cut away the plastic waveguide edge myself, rather than attempting fancy router work.

To reinstall the plastic trim, shave it down where necessary to clear the waveguide rim on each side (Figure I). Nail in the top, bottom, and one side piece using ¼" brads, and screw in the other side so it can be removed for service.

Solder the woofer wires to their terminals, observing polarity. Screw the woofer down and replace the grille (Figures J and K). That's it!

Options

Here are some additional restoration options, mostly cosmetic, that won't affect the speakers' sound:

» For a homogenous appearance, paint the cabinet back, woofer frame, baffle, and screws all black. You'll need ½ pint of interior latex trim paint.

» To mount the woofers more securely, drill out their screw holes to ⁷⁄₃₂" and press in eight 8-32 T-nuts.
» Make new grilles using the Large Corner Grill Frame Kit and an extra set of corners (Parts Express #260-344 and #260-345). Choose thin, stretchy fabric, like polyester jersey fabric (1yd). Attach the grille with velcro tape, and the fabric with foam tape.

Ross Hershberger posts on Audiokarma (audiokarma.org) as Bauhausler. He's been interfering with electronic stuff of all kinds for 40 years.

Make: TIPS

Cross-Grain Sanding
You should always sand with the grain, right? Well, not necessarily. Take spruce or cedar, for example. These woods have such soft grain between the hard winter growth rings that particles of abrasive from the sandpaper that break off just roll around in the soft grain, scarring it.

If you sand across the grain, you'll cut the hard and soft grain equally. And, as you get to the finer grits, the sanding scratches become so small that they don't show under even the most transparent finishes.
—*Frank Ford*, frets.com/homeshoptech

Find more tools-n-tips at makezine.com/tnt.

RETRO WIRELESS HANDSET

Adding Bluetooth to an old phone.
By Jeff Keyzer

Like it or not, Bluetooth headsets have become a part of our lives. Formerly worn only by stuffy corporate execs and the social elite, these devices have become ubiquitous, thanks to dropping prices and hands-free driving laws. A quick search online shows that a new Bluetooth headset can be purchased for less than $20. This is great news for makers because now we can use them in all sorts of projects!

My wife gave me her old Motorola HS820 Bluetooth headset when she upgraded to a newer model. Of course, I immediately tore it open. Inside, I found a small printed circuit board, a lithium polymer (LiPo) battery, an electret microphone, and a tiny speaker.

I came across a vintage ITT telephone handset on a recent trip to Weird Stuff Warehouse in Sunnyvale, Calif. The instant I saw it, I knew exactly what to do:

Photography by Jeff Keyzer

MATERIALS AND TOOLS

Bluetooth headset An older, larger model is better since it will likely be easier to hack.
Vintage telephone handset The kind with round, removable ear and mouthpieces.
Superbright 5mm blue LED All Electronics part **#LED-122,** allelectronics.com
Panel-mount LED holder All Electronics **#HLED-3**
Normally open panel-mount push-button switch All Electronics **#MPB-1**
Coaxial panel-mount DC power jack; matching plug
Fine-gauge insulated stranded wire
1kΩ resistor You might need some other value, so have an assortment on hand.
Soldering iron
Hot glue gun
Multimeter with continuity tester
Wire cutter/stripper
Handheld drill and drill bits
Small screwdriver

stick the guts of the Bluetooth headset inside and create a retro Bluetooth handset!

1. Open and prepare the Bluetooth headset.

Remove the screws on the back of the headset and pry the case open, exposing the printed circuit board (PCB) and wiring inside. Make a diagram showing where the wires for the speaker, microphone, and battery connect to the PCB (Figure A).

The polarity of the battery is important, so be sure to mark where the red (positive) wire goes. Unsolder all the wires, starting with the battery. Make sure the battery wires don't short together. I used a piece of tape to keep them safely apart.

NOTE: Not all Bluetooth headsets are the same. Yours may differ slightly from that shown but will likely have all of the same parts.

2. Remove the microphone and speaker from the handset.

Unscrew the handset's round caps (Figure B). The microphone is held against spring terminals and tends to jump out, so be careful it doesn't roll away. The speaker and the plastic microphone holder should pull out easily. Remove all the original wiring from the handset and set it aside.

NOTE: You can reuse the old wiring for the hack, but if you do, make sure to untwist the ends and cut away the flexible cotton fibers inside before you solder, or you'll end up making a mess.

3. Install the push button and LED.

Choose a location for the push button and drill a hole slightly larger than the threads. Solder a few inches of wire to each terminal and install the push button, securing it with the supplied nut. My headset had a micro switch in the center of the PCB to answer calls and toggle power. Yours will likely have a button somewhere that does the same thing. Desolder the button from the PCB and solder the wires for the new push button in its place (Figure C).

My headset also had volume control buttons, but I left those in place because I can control the volume from my cellphone instead.

To add the LED, drill a hole in the handset just large enough for the LED holder. Solder wires to the legs of the LED and install the LED in the holder. Push the

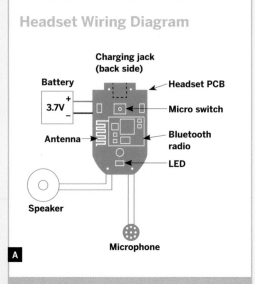

Headset Wiring Diagram

Charging jack (back side)
Battery
Headset PCB
3.7V
Micro switch
Antenna
Bluetooth radio
LED
Speaker
Microphone

A

Fig. A: Here's how my Bluetooth headset was wired. Yours may vary — so draw your own diagram, identifying these important parts.

entire assembly into the hole you just drilled — it should be a tight fit. Desolder the surface-mount LED from the headset PCB and carefully solder the 2 wires for the new LED to the exposed pads. The polarity matters, so if the LED fails to light up later when testing the headset, try reversing the wires.

4. Install the microphone and speaker.

Attach a few inches of wire to each terminal of the handset speaker and feed them through the handle. Carefully solder the wires to the speaker connections on the headset PCB (Figure D), using your diagram as a reference.

I was able to get the carbon microphone on the handset to work by adding a 1K series resistor and wiring it to the microphone terminals on the PCB. Without the resistor, my voice was too loud and distorted. You may have to experiment to find a resistor value that works best for you. For better audio quality, you can install an electret microphone (RadioShack #270-092) instead.

5. Install the charging jack.

Remove the original charging connector from the headset. To find the ground pin, test for continuity

Fig. B: The telephone handset, disassembled.
Fig. C: The headset's PCB, showing the wiring for the push button (center) and LED (bottom right).
Fig. D: The reverse side of the PCB, showing the battery and wires for the charger (upper right), microphone (lower left), and speaker (lower right).
Fig. E: The charging jack and flat washer spacer installed.
Fig. F: The modified charger with coaxial power plug.

between each pin and the negative battery terminal on the PCB. Mount the coaxial power jack on the handset, taking advantage of the round hole left behind by the cord. I added a washer so the jack would fill the relatively large opening (Figure E). Solder wires from the power jack to the charging terminals on the PCB, making the center pin the positive wire.

6. Modify the charger.

The charger used a proprietary 3-pin connector to connect to the headset, so I cut it off. Only 2 of the pins are actually used for charging, so I replaced the connector with a more common coaxial plug (Figure F) to match the jack on the phone. Plug in the charger and use the multimeter to determine the polarity of the wires. Then connect the positive wire to the center of the plug, and the negative wire to the outside terminal.

7. Assemble the handset.

Reinstall the headset battery, making sure to solder the wires with the right polarity. Use hot glue to secure the battery to the PCB. Put a dab of hot glue over each wire's soldered connection to the PCB. This gives the wires some strain relief and keeps

them from ripping the traces off the PCB.

Insert the headset PCB into the handset behind the microphone holder. Install the speaker, microphone holder, and microphone. I had to remove a plastic tab from the back of the microphone holder to keep it from hitting the charging jack inside the handset. Screw the caps back onto the handset.

Follow the headset manufacturer's instructions to turn on your new retro wireless handset and pair it with your cellphone. Call a friend and test it out. If everything works, admire your finished handset.

You're done!

Jeff Keyzer is an electrical engineer who lives in San Francisco and hacks cars and microcontrollers. Visit his blog to see more projects like this one at mightyohm.com.

Drilling on a Round Object
Next time you need to drill a hole in something round, file a little flat spot so the drill can get a good start without skating off to the side.
—Frank Ford, frets.com/homeshoptech

Find more tools-n-tips at makezine.com/tnt.

Property Condemned

The Scenario: You belong to an amateur but dedicated group of "urban archaeologists" who have made it their mission to document the vanishing details of your city's history. You notice while driving home late one afternoon that a classic building your group has often admired is now fenced off and slated for demolition. You decide that a quick in and out — a few notes and some photographs — will serve your mission well.

Parking out of sight, you ignore the "No Trespassing" signs, load your minimal gear in your coat pockets, and easily scale the fence to gain access to the building. Guided by your small but powerful LED flashlight, you wend your way up the creaky wooden staircases, snapping pictures all the while with a compact digital camera.

The Challenge: As you cross one of the upper stories, you suddenly crash through some floorboards and find yourself wedged in the hole tight, up to your armpits. What's more, the drop has forced the splintered edges of the broken floorboards downward along your body. So even though your arms and shoulders seem OK, if you try and lift yourself out, the boards act like a giant, wicked Chinese finger puzzle and just dig in deeper, wedging you in even tighter.

You don't know how far the drop is below you, and the hole is so tight that it's doubtful you could get your shoulders through it anyway. Hanging there all night in hopes that the wrecking crew might find you in the morning might be an option — except that the massive rats could probably put a serious dent in you by then. So, intrepid archaeologist that you are, how are you going to get yourself out of this one?

What You've Got: You have a flashlight and camera, both of which are within reach. You've also got a working cellphone and a Swiss Army knife (or Leatherman tool) — except *both of those* are in your coat pockets, which are below the hole. And, is it just your imagination, or is that rat in the corner actually smiling at you?

Send a detailed description of your MakeShift solution with sketches and/or photos to makeshift@makezine.com by Feb. 19, 2010. If duplicate solutions are submitted, the winner will be determined by the quality of the explanation and presentation. The most plausible and most creative solutions will each win a MAKE T-shirt and a *MAKE Pocket Ref*. Think positive and include your shirt size and contact information with your solution. Good luck! For readers' solutions to previous MakeShift challenges, visit makezine.com/makeshift.

Lee David Zlotoff is a writer/producer/director among whose numerous credits is creator of *MacGyver*. He is also president of Custom Image Concepts (customimageconcepts.com).

Motion-memory toys, best holiday kits for kids and grownups, cool new tools, books, sites, and the ultimate Dremel.

TOOLBOX

Topobo
$149 for 50-piece set, $249 for 100-piece set
MAKE readers: Use promo code "MAKE" to get 50% off!
topobo.com

My living room looks like Darwin's laboratory. There's a multicolored, three-legged creature writhing on the floor. My daughter pulls off one of its legs, and connects it to its face. My son teaches it to walk. Much better — it's really moving now! This is how you play and learn with Topobo.

This brainchild of the MIT Media Lab's Hayes Raffle and Amanda Parkes is part construction toy, part kinetic memory robot. By combining passive construction pieces with active motor/brain hubs (learning servos) you can spend hours creating unlikely creatures with even more unlikely forms of locomotion.

My kids (6 and 3) immediately understood how to

record and play back the motion of their creations; press a button, perform a few poses, press the button again, and it's alive! They were delighted to see the colorful creatures twitching, creeping, and lurching around.

Our only complaints: the Lego Technic-compatible peg connectors can be hard to remove, and we'd prefer battery power to being tethered to the wall with the power supply wire.

While I wished the motion memory could be saved between sessions, my kids had no interest in reproducing past creations. Topobo is all about exploring novel methods of movement, something my kids figured out before I did.

—*John Edgar Park*

» **Want more? Check out our searchable online database of tips and tools at** makezine.com/tnt.
Have a tool worth keeping in your toolbox? Let us know at toolbox@makezine.com.

Motor Bicycling

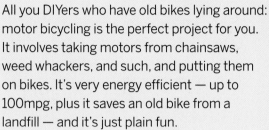

motorbicycling.com

All you DIYers who have old bikes lying around: motor bicycling is the perfect project for you. It involves taking motors from chainsaws, weed whackers, and such, and putting them on bikes. It's very energy efficient — up to 100mpg, plus it saves an old bike from a landfill — and it's just plain fun.

Motorbicycling.com is a friendly forum where all newbies are welcomed with open arms. You can get expert advice and safety tips, and follow other people's projects, or you can make your own thread, post pictures, and write reviews of your own motorized bike. And when I say expert advice, I mean it. Members with a lot of experience share with you to keep frustration at a minimum on your bike build or kit. Whether you use the friction, direct, or pusher methods (to learn what these are, check out the site), this is the ultimate resource for DIY motor bikers.   —Sam Fraley

Dremel 4000 Rotary Tool Kit

$79–$99 dremel.com

High-speed rotary tools make small cutting and grinding jobs easy, and can punch through hard materials you'd botch with a slower-spinning drill. Many makers can't imagine a workshop without one.

To overhaul their top-of-the-line tool, Dremel quizzed customers and engineered some of the most-requested improvements. The new Dremel 4000 senses loads to maintain motor speed. For better control, it's slimmed down, with a pencil-grip nose, 360° rubber grip, literal-reading speed dial (5,000rpm–35,000rpm), and a separate power switch to save your speed setting. It's got a bigger fan motor, replaceable motor brushes, a five-year warranty, and two smart new attachments: a "detailer's grip" to balance the tool like a stylus, and a sanding/grinding guide to help it follow edges.

The tool's speed never wavered when I leaned on it to grind down a misshapen PVC fitting, and its balance and control are very good. Upshot: a real upgrade of a classic tool. And I like that Dremel listened to makers to get it dialed in.   —Keith Hammond

Fire Pit Kits

$95 and up formandreform.com

Blacksmith/sculptor John Sarriugarte of Form and Reform makes magical propane-powered fire pits. As he puts it, "Not only do they give off warmth but you can entertain your guest with drawings in the sand that burn where you draw!" You can buy one ready-made, or pick up one of his fire pit kits to make your own.

—*Arwen O'Reilly Griffith*

3D Animal Sculpture Kits

$15 and up 323d.com

These kits debuted in a limited edition at the *Howtoons* booth at Maker Faire, and now they're available on the web. Shipped to you in laser-cut, flat-pack cardboard, they layer into incredible creatures with some glue and a few hours of work. Try out the easy hippo, or devote an entire day to the horizontally built frog. **—AOG**

YurtDome Kits

$290+ shelter-systems.com

I've been playing around with the idea of building a dome for years, and when I was looking for a quick and relatively inexpensive place to put my workshop for the winter, this kit from Shelter Systems seemed like just the ticket.

Shelter Systems is still based in Santa Cruz, Calif., and using the same non-puncturing "grip clip" technology they had back in the 80s to make mod-friendly yurt-dome hybrid structures (panels are shingled together and easily replaceable).

These things are adaptable for snow and wood-stoves, and they're great for greenhouses (they also come with translucent panels). The tent was surprisingly easy and fun to set up (we actually had to fight off volunteers who wanted to help), and we got it standing in about 30 minutes. Let it rain! **—Meara O'Reilly**

Metabots Kit

$15 stores.microrobotestore.com

If doing a lot with a little is an important part of modern design, it's easy to see why the Metabots toy robot building kits by Korean manufacturer EnjoyMobil have found their way into the Museum of Modern Art's toy collection.

What you get in the box is nothing more than a couple sheets of die-cut foamcore, a dozen injection-molded plastic ball joints, and some instructions. But what you end up with, after about 90 minutes of punching out and slotting together, is one of the coolest posable robot toys on the market today — certainly way cooler than anything else in its price range. As icing on the cake, each of the seven Metabot models is also available in an all-white "prototype" version to encourage user customization. —*Sean Ragan*

20-Piece Bit Driver Kit

$15 ifixit.com

I always resist buying kitted tools because they never seem high-quality, but this kit is exactly what I need when my electronics break. (iFixit has replacement parts and free tutorials, too, so you don't have to ditch that cracked iPhone or shell out to have the Apple Store fix it.) The driver kit has all the bits you'd want, in nice sizes, and with a convenient case — a comprehensive toolkit for the electronics hacker.

It's the sort of thing you probably wouldn't buy for yourself, but would be stoked to get as a gift. So buy one for a worthy friend, or just make sure to suggest strongly to your better half that it should show up in your stocking.   —Saul Griffith

NerdKits

$80 nerdkits.com

Ever since I got into robotics, I've wanted to know more about electronics and programming. I wanted a good kit with a microcontroller, but the kits I found were either too advanced or too dumbed-down without many interesting components.

I finally found a good kit. A USB NerdKit comes with a lot of great stuff, including an ATmega168 MCU, a buzzer, a 4×20-character LCD display, three different colors of LEDs, a temperature sensor, switches, and a potentiometer. It even comes with a great guide and programmer! The guide starts at the basics and works its way up to more advanced topics. The programmer comes with multiple languages and sample codes.

The NerdKits support team is quick and helpful. There's a great forum and even video tutorials that explain how the programs and circuits work. I would definitely recommend a NerdKit!   —Dylan Kirdahy

Sea Perch ROV

About $75 seaperch.mit.edu

My favorite scene from Maker Faire 2009 was two girls laughing as they remote-piloted tiny submersibles around a big tank of water. The Sea Perch ROV was developed by MIT Sea Grant and the U.S. Navy to spark students' interest in robotics, engineering, and marine sciences. Obviously kids loved driving the Sea Perch. Could it really be that easy for kids to make?

I decided to build one, and yes, it's that easy. The parts are standard stuff from hobby shops, hardware stores, and RadioShack or Jameco. The free, downloadable instruction manual is classroom-clear, and the vehicle design is simple and forgiving, with DC motors potted in wax, and a frame made of slip-fit PVC pipe that allows infinite adjustments. It's the first project I've soldered, and it was a piece of cake.

Bagged kits are provided for teacher trainings (sign up at seaperch.mit.edu), but anyone can download the manual and parts list and make their own kit. The website also has video instructions, K–12 and college curriculum ideas, an international data bank where students can upload data, and tips on adding cameras and sensors.   —KH

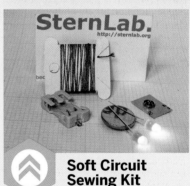

Soft Circuit Sewing Kit

$15 bekathwia.etsy.com

Not to toot MAKE and CRAFT blogger Becky Stern's horn, but she's put together a fantastic kit for augmenting a favorite garment or a needlework project with LEDs. The kit is based on the demo she gives at Maker Faire as an introduction to sewing a soft circuit. There's a free video tutorial and step-by-step instructions if you're not used to working with soft circuits, and the kit includes everything you need to get started; just be sure to tell her what color LEDs you want!   —AOG

Ice Tube Clock Kit

Vacuum Fluorescent Display Clock Kit
$70 **Maker Shed #MKAD16**

Somewhere at the junction of modern open source hardware and early-80s Russia lives this beautiful new DIY kit. The centerpiece is a Russian-made, 9-digit, vacuum fluorescent display (VFD). Remember those bright green VCR and boombox displays? Yep, those were most likely VFDs.

The power source, case, VFD tube, and electrical components are ready to be soldered together right out of the box. The kit features an alarm with snooze function and adjustable brightness — perfect for your nightstand.

The kit went together easily, but I do suggest having some soldering skills already. There are a lot of parts, but none of them are particularly difficult to solder, and the online instructions are very clear. The case goes together easily with only a small Phillips-head screwdriver.

Give it as a gift, or keep it for yourself; either way, the end result is a clock that anyone would be happy to have. Don't forget, since it's open source, it's completely hackable, too.

LED Clock Kit

$35 **Maker Shed #MLEDB1, MLEDB2**

The LED Clock Kit is another great no-solder project from the Maker Shed. No printed circuit boards — just twist the wires together and enjoy the results! The finished clock measures 9"×5" and you can change the brightness and program the time with the push of a button. It comes with blue or red LEDs.

LED Art Kit

$20 **Maker Shed #MKKM2**

Create your own unique LED light show with this easy-to-assemble, no-solder kit. The only tool needed is a pair of pliers for crimping the wire connectors. The RGB (red-green-blue) LEDs create a slow, ever-changing multicolored glow that's perfect for setting the "mood" of any room. Makes a great nightlight, too!

Snap Circuits Jr. Kit

$30 **Maker Shed #MKEL6**

Learn the basics of electronics through more than 100 exciting projects in this award-winning kit. Start by building simple switches and lights, and later move on to more complex projects like alarm systems and circuits that make music, all with no soldering.

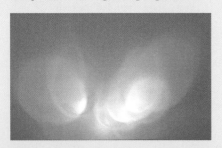

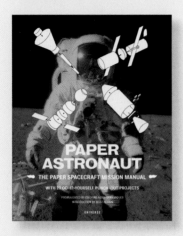

« Paper Moon

Paper Astronaut: The Paper Spacecraft Mission Manual by Juliette Cezzar
$28 **Universe Publishing**

At first glance, it's clear that this is no ordinary papercraft book. The visually stunning layout and amazing archival photos make it a delight to page through. More than half of the large book is taken up with exploded color-coded diagrams, striking full-page photographs, and detailed descriptions of the 20 spacecraft that the brave can then assemble out of the templates in the second half of the book.

Be forewarned, these are no simple paper models. The book's imagery will appeal to enthusiasts of all ages, but the papercraft work is not suitable for young children. Still, the photos are worth the price alone as *Paper Astronaut* provides an inspirational and hands-on look into spacecraft and how they work. —*Bruce Stewart*

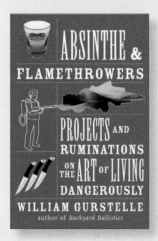

« Living on the Edge

Absinthe and Flamethrowers: Projects and Ruminations on the Art of Living Dangerously by William Gurstelle
$17 **Chicago Review Press (Available at** makershed.com**)**

Society today is less interested in nurturing risk-takers than it is in being "safe and sane." Stuff that you did as a kid — chemistry sets, model rockets — has now been made so vanilla it's boring. Not so with Bill Gurstelle's latest book, in which he asks you first to answer: Are you a Big-T person or a little-t, a thrill seeker or a meek milquetoast?

Most of us, myself included, fall somewhere in between. But you don't have to be a Big-T to enjoy this book, which is full of projects and tips ranging from how to buy quality absinthe (already ordered some) to making a rocket from scratch, and culminates with building your own flamethrower (my wife said no!). Gurstelle takes on everything with a perfect blend of history and how-tos. As educational as it is entertaining, this book will guide you toward getting your man-card. Y'know, if it's OK with your wife. —*Rob Bullington*

« Real Earth

Whole Earth Discipline: An Ecopragmatist Manifesto by Stewart Brand
$26 **Viking Adult**

Stewart Brand's "Ecopragmatist Manifesto" is a tour de force of persuasion, using the urgency of climate change to re-examine environmental orthodoxy. His conclusion: there is no "natural." Cities are green, nuclear power is green, genetically modified crops are green. "Never mind terraforming Mars," he says. "We've already terraformed Earth." We're just doing it badly.

I had already heard the arguments for cities and nuclear power; what was most revelatory to me was his argument for genetically modified foods. Even if you disagree with Brand's conclusions, though, you will be far smarter by the time you finish the book. It is a backstage tour of a remarkable mind, 50-plus years of reading, and the equivalent of decades of TED talks, compressed into 300 pages. Read it, pass it on, act on it. —*Tim O'Reilly*

Archos 5 Media Player

$280 archos.com

I have owned quite a few media players, but none has impressed me as much as this one. It's equipped with a 4.8" touchscreen and wi-fi, and even runs Linux. It has an ARM Cortex processor that runs at 600MHz, so everything runs smoothly, and supports a multitude of media codecs, so I never worry if files are compatible.

It's a large device, but the screen size and storage capacity justify it. I'm able to get about 5 hours of video out of it, which is impressive. When I'm waiting for the bus, or doing anything where I have some free time, the ability to have 60GB worth of music and video with me at any time is outstanding.   *—Hayden Lutek*

X-Bench Portable Workstation

$169 skiltools.com

Much to my wife's chagrin, my DIY and technology projects have often found themselves firmly entrenched upon our dining room table. While it might be too late to save my dining room set, it's not too late for yours, thanks to the X-Bench.

This portable workstation can be set up virtually anyplace that will accommodate a folding utility table. The workstation provides a sturdy MDF work surface measuring 53"×23". The left side accommodates Skil's "universal insert plate system" which lets users easily convert the unit into a scroll saw, drill press, or sanding station. A miter gauge slot is located to the left of the plate.

Meanwhile, the right side of the table can be extended to allow for a center cut channel. Both surfaces incorporate numerous peg holes intended to accept the included wedge-shaped clamps. Ruler markings are also provided, as is a bump-off power switch.

After you've completed your project using the X-Bench, all that's left to do is store it. Your significant other will be glad you did.   *—Joseph Pasquini*

Automatically Adjusting Wire Stripper

$17 makezine.com/go/wirestripper

I bought two of these from Micro-Mark several years ago when they were priced at $30; now they're $17. Even so, I would gladly pay $30 again. The jaws automatically adjust to remove insulation from any size wire from #26 to #10 AWG. All you have to do is squeeze the handles. With its integral wire cutter and terminal crimpers, this tool can handle many light electrical jobs by itself, which makes it a great choice if you have limited space or weight capacity for tools. After my multimeter, it's always the first tool I grab out of my toolbox when I'm doing electrical work. It's a joy to use, both because it works so well and because the mechanics of the stripping head are interesting to watch.

There's an adjustment dial on one jaw that must sometimes be tweaked, as in the case of stripping especially delicate wire, but 95% of the time the automatic wire strippers do their job perfectly and without complaint. Highly recommended.   **—SR**

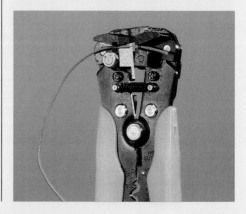

Tricks of the Trade By Tim Lillis

Puttin' on the bricks.

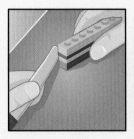
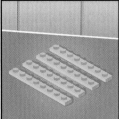

Are you sick of having piles of unorganized Lego bricks, some randomly placed, some stuck together forever? Evil Mad Scientist Laboratories has you covered.

This method works best with 1×6, 1×8, and 1×10 plates. Start by laying a few plates parallel until you have roughly a square shape.

Now lay the same amount on top at a perpendicular angle. You can also take this opportunity to organize your plates by color.

Stack until you have them all stored! For more useful Lego organization tricks, check out their website at evilmadscientist.com/go/efficientlego.

Have a trick of the trade? Send it to tricks@makezine.com.

Retro Word Pro

Pilot Varsity Fountain Pen

$3–$18 (pack) Office supply stores

The Pilot Varsity is sold as a disposable fountain pen. However, with a little careful work, you can remove the nib section with a pair of pliers and refill the pen with your favorite ink.

The Varsity is an amazing little pen — it always writes, even after sitting in a drawer for months. That's more than I can say for some of the expensive pens I own. I love to give them as gifts to people who do a lot of writing. Try one and see for yourself.

—Anton Ninno

Hyperlinked Notebook

MoinMoin Desktop Edition Personal Wiki

Free moinmo.in/DesktopEdition

Every maker needs to work out ideas, and while notebooks and text files are great, it's hard to show links between ideas. This is where a personal wiki can be useful. The MoinMoin wiki package is available in a Desktop Edition that will run on Windows, Mac, and other Unix and Linux flavors. The best part is that I don't have to install or configure any additional software. (Windows users will have to install Python.)

Now I've got my own wiki that I can use to work on my ideas. I've got all the features I'm used to, like the ability to edit any page and change logs for everything I do. I'm currently using it to collect ideas I get for game projects and for the web app I'd like to build in the future. Any maker who wants a hyperlinked notebook should invest some time in MoinMoin Desktop Edition. **—David Delony**

Rob Bullington is online manager of the Maker Shed store and handles marketing and events for MAKE.

David Delony is a technology enthusiast and writer based in the Bay Area.

Sam Fraley, age 11, lives in St. Louis, Mo., and is an avid garage hacker. sl.fraley@gmail.com

Saul Griffith is a co-author of *Howtoons* and a MacArthur Fellow.

Dylan Kirdahy is a 12-year-old robotics enthusiast.

Hayden Lutek is obsessed with tech, from Apple to ZFS.

Anton Ninno is a K–8 technology teacher at the Southside Academy Charter School in Syracuse, N.Y.

Meara O'Reilly is an intern at MAKE.

Tim O'Reilly is the founder and CEO of O'Reilly Media.

Joseph Pasquini is an avid amateur radio operator and shortwave listener.

Sean Michael Ragan's ancestors have been using tools for 5,000 generations.

Bruce Stewart is a freelance technology writer and infrequent contributor to *Wired*'s geekdad.com.

Have you used something worth keeping in your toolbox? Let us know at toolbox@makezine.com.

USE A PAPER PLATE TO CREATE.

CUT 8 SLICES IN THE INNER CIRCLE.

FOLD OUT EVERY OTHER SLICE.

LIKE SO!

SWAP OUT THE PAPER PLATE FOR A PIECE OF CARDBOARD. HOW BIG CAN YOU GO?!

THE FOLD-OUT WINGS PROVIDE THE SURFACE FOR THE WIND TO PROPEL THE PLATE. THE GYROSCOPIC EFFECT OF ROTATION KEEPS IT BALANCED. THIS IS A SELF-BALANCING PAPER PLATE !

GO BABY GO!

REMAKING HISTORY
By William Gurstelle

Humphry Davy and the Arc Light

» Thomas Edison did not invent the first electric light.* More than 70 years before Edison's 1879 incandescent lamp patent, the English scientist Humphry Davy developed a technique for producing controlled light from electricity.

Sir Humphry Davy (1778–1829) was one of the giants of 19th-century science. A fellow of the prestigious Royal Society, Davy is credited with discovering, and first isolating, elemental sodium, potassium, calcium, magnesium, boron, barium, and strontium. A pioneer in electrochemistry, he also developed the first medical use of nitrous oxide and invented the miner's safety lamp. The safety lamp alone is directly responsible for saving hundreds, if not thousands, of miners' lives.

But it is his invention of the arc lamp for which we remember him here. Davy's artificial electric light consisted of two carbon rods, made from wood charcoal, connected to the terminals of an enormous collection of voltaic cells. (In Davy's day, thousands of cells, similar to modern chemical batteries, had to be wired together in series to produce the voltage required to strike an arc between the carbon electrodes.) When Davy closed the switch connecting the battery to the electrodes, electricity jumped between the carbon tips. The result — a continuous, glaring, lucent dot of white heat — was so bright that it was dangerous to look at for more than a split second.

While making an arc light isn't terribly complicated, the arc's underlying physical processes are indeed complex. Although normally a nonconductor, carbon will conduct electricity in certain circumstances. The graphite rods used in arc lights conduct electricity, albeit grudgingly, if enough electrical potential is applied. At high voltage levels, the rod tips become white-hot, and carbon particles break away from the main body of the rod. Within the resulting particulate mist, small bits glow white with heat and jump across the spark gap between electrodes. This produces the incandescent arch of light known as an *electric arc*.

"The Dazzling Splendor," as Davy called it, was a tricky beast to control. After the initial sparks

BRILLIANT MISTAKES: Humphry Davy, chemist, inventor, and philosopher: "I have learned more from my failures than from my successes."

appeared between the electrode tips, Davy had to separate the carbon electrodes slightly and carefully in order to sustain the continuous, bright arc of electricity. Once that was accomplished, he found the device could sustain the arc for long periods, even as the carbon rods were consumed in the heat of the process.

Davy's arc lamp of 1807 was not economically practical until the cost of producing a 50V-or-so power supply became reasonable. This didn't occur until the mid-1870s, with the introduction and commercialization of the electrical dynamo. But as soon as that happened, arc lights were everywhere, archetypically in searchlights, as well as in lighthouses, streetlights, movie sets, and movie projectors.

It took a lot of juice to run a searchlight. To maintain its arc, a 60-inch-diameter World War II vintage carbon arc searchlight drew about 150 amps at 78 volts, which is roughly equivalent to a 12,000-watt light bulb. A lot of power, yes, but it could light up an airplane 5 miles away.

Perhaps the largest carbon arc lamp ever made was the 80-inch-diameter monster searchlight built at the turn of the 20th century by General Electric. It lit the grounds of the 1904 St. Louis World's Fair with its billion-candlepower arc light.

This month's Remaking History project shows how to make a smaller version of the dazzling splendor in your workshop without too much trouble.

It seems that Edison may deserve credit for the first electric light bulb, but even that is controversial.

Image from National Portrait Gallery, London

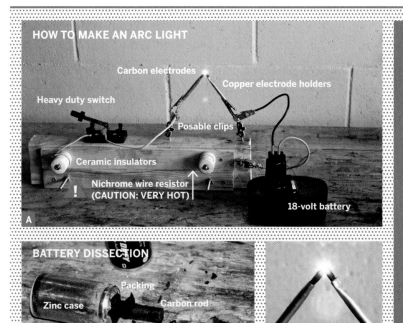

HOW TO MAKE AN ARC LIGHT

Carbon electrodes

Copper electrode holders

Heavy duty switch

Posable clips

Ceramic insulators

Nichrome wire resistor
(CAUTION: VERY HOT)

18-volt battery

A

BATTERY DISSECTION

Packing

Zinc case

Carbon rod

B

C

THE DAZZLING SPLENDOR: You can build Sir Davy's world-changing 1807 invention out of simple materials such as copper and carbon (Figure A), powered by a 12V or 18V battery instead of Davy's primitive voltaic cells. He used charcoal, but nowadays it's easiest to get your carbon rods straight out of non-alkaline batteries (Figure B). When electricity is applied, small bits of hot carbon jump across the gap, producing a dazzling incandescent arc (Figure C).

 SAFETY CAUTIONS

Please read before beginning the project.

» The arc light produces *strong ultraviolet light* that can damage skin and eyes. For safety, use arc welding style eye and skin protection including gloves, long sleeves, and a helmet with #7 shade or darker when using the arc light.

» The Nichrome wire and copper electrode holders get *extremely hot*. Be very careful around them! There is little to no shock hazard associated with 12- and 18-volt batteries.

» This is a demonstration device only and should only be operated intermittently and for brief intervals. *Running the arc lamp for too long can damage your battery or battery charger.* If using a battery charger, check the ammeter on your charger to make sure the circuit is not shorted. If it is, or nearly is, use a longer Nichrome wire.

MATERIALS AND TOOLS

12-volt battery charger with ammeter, or an 18-volt battery from, say, a portable power drill
Carbon rod, about ¼" diameter, ½" lengths (2) The easiest way to obtain a pure carbon rod is to cut open a regular non-alkaline AA, C, or D cell battery with a Dremel tool or hacksaw. Such batteries are usually labeled "heavy duty" or "non-alkaline." Cut off the top and carefully remove the carbon rod from the black, greasy packing that surrounds it. The packing material will stain hands, clothes, and work surfaces, so wear rubber gloves and cover surfaces with newspaper.
Miscellaneous wood pieces
20- to 24-gauge Nichrome wire, 2'
Porcelain insulators (2) Electric fence insulators work well.
Nuts and bolts to mount insulators to wood base
Heavy-duty flexible stranded lamp cord
Small copper spring clamps (2)
¼"-diameter copper tubing, 1½" lengths (2)
Heavy-duty on/off switch aka knife switch
Posable alligator clips I used clips from a discarded "third hand" tool, but you could easily make them by soldering an alligator clip to a stout copper wire.
Miscellaneous wood screws
Wire stripper/crimper
Needlenose pliers
Screwdriver
Sandpaper
Gloves
Eye protection such as an arc welding helmet

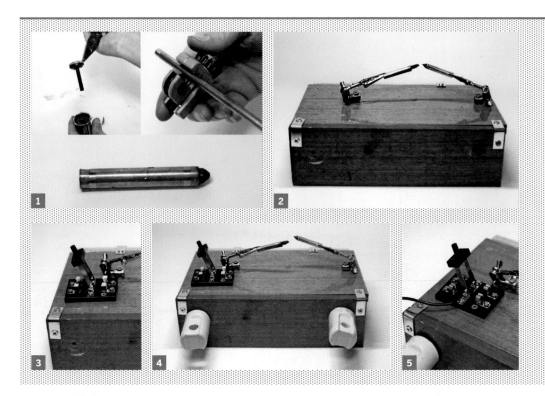

Making a Davy Carbon Arc Light

1. Carefully clean the 2 carbon electrodes. Sand them until they fit snugly into the ¼"-diameter copper tubes. Crimp each carbon rod into place in a copper tube. Sand the protruding end into a point.

2. Mount the posable clips to a wood frame as shown. Position the electrodes so the carbon points are just barely touching.

3. Mount the on/off switch as shown.

4. Mount the ceramic insulators as shown in the photo, approximately 10" apart.

5. Wire the circuit. The circuit is very simple. Electricity from the battery goes to the first electrode holder, through the carbon electrode, across a very small spark gap to the second electrode. From there, the electricity goes through the main on/off switch and then across a length of Nichrome wire before entering the opposite pole of the battery or battery charger.

6. Close the switch and carefully adjust the spark gap until a bright white light is obtained.

7. Once a bright arc is struck and maintained, you can optimize the light output of the system by making the Nichrome resistor wire longer or shorter.

Operation

Every homemade arc light is a bit different. Make adjustments as necessary. The spacing of the electrode gap is critical, so take your time adjusting it in order to obtain the best arc light. Too much or too little contact will result in no arc light.

Your battery will be damaged if the circuit is run without adequate load. The Nichrome wire provides just enough resistance to prevent the battery from shorting.

You'll have to adjust the length of the Nichrome wire for best performance; if it's too short, it will quickly burn up, but if it's too long, the arc light will be dim. Find the correct length through trial and error.

William Gurstelle is a contributing editor of MAKE magazine.

Photography by Ed Troxell

AHA! | Puzzle This By Michael H. Pryor

MAKE's favorite puzzles. (When you're ready to check your answers, visit makezine.com/20/aha.)

Tanks A Lot

Four kids started aquariums in their houses. They each had one fish and one invertebrate in their tanks. Unfortunately, they also each had a different problem with their tanks. One tank had lighting issues, another had pH problems, one couldn't maintain the right temperature, and the other one had an algae outbreak. The kids each fed their tanks at different times of day (morning, afternoon, evening, and night).

Figure out who owned which tank and had which problem; the fish and the invertebrate that lived there; and the time of day that the owners fed their tanks.

1. The aquarium that was fed in the evening had an algae bloom problem. Victor did not feed his tank in the morning.

2. The tank with temperature problems had an urchin in it. Ruth did not feed her tank in the evening.

3. The lionfish did not have a pH problem in its tank. The clownfish tank had lighting issues.

4. The anemone's tank was not fed at night.

5. Victor, who did not have a clownfish, had a clam in his tank. Aviva's scooter blenny was not in the tank with the algae problem.

6. Michael, who did not have a clownfish, fed his tank in the afternoon. The blue tang and the shrimp lived in the same tank, but they did not eat in the morning.

Michael Pryor is the co-founder and president of Fog Creek Software. He runs a technical interview site at techinterview.org.

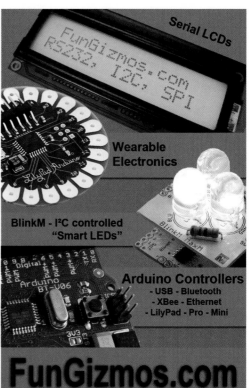

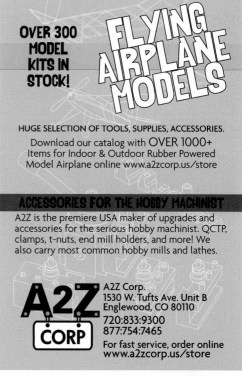

$1.00
Church key from
a country store.

At the checkout counter of an old country store I saw a bowl of stamped metal bottle openers. It was marked with masking tape: Church Keys — $1.00 each!

This got me wondering what sort of bottle openers I could make for just half that price, using real money.

I made the first one by pinching 2 quarters in a vise and and bending them with a hardwood mallet. I fastened them together with a single steel pop rivet.

I fashioned the next one from a single Kennedy half-dollar. This opener is designed to be stamped from an industrial punch press — but that would require a custom tool and die. To make the proto-type, I used a tiny jeweler's bit to drill a 0.5mm hole

↑ $0.50
Bottle openers made
from coins.

through the coin. Then I sawed out the shape of the tooth using a jeweler's saw with a Size 0 blade. The rest was easy: careful bending and filing with the coin held in a vise.

My favorite opener required no bending. I made 3 cuts in a quarter using a hacksaw. Then I made matching cuts in 3 other coins using different saw blades so that the kerf of the cuts matched the thickness of the coins. I snapped them all together, brushed on some soldering flux, heated them softly with a MAPP gas torch, and added a dab of "silver solder" braze.

A quarter, 2 dimes, and a nickel: still just 50 cents, but this one works like a champ and looks like a million bucks!

MAKER'S CALENDAR
Compiled by William Gurstelle

Our favorite events from around the world.

Consumer Electronics Show 2010
Jan. 7–10, Las Vegas
The premiere show for electronics buffs, the gigantic CES is the place to see what kinds of new electronics will soon be available for hacking and modifying. Over 2,700 vendors will show off things electronic, ranging from smartphones to 3D projectors. cesweb.org

>> DECEMBER

>> The Royal Institution Christmas Lectures
Dec. 5–16, London, England
This extraordinary series began in 1825 with Michael Faraday's presentation of lectures for young people. Luminaries since then include David Attenborough, Carl Sagan, and Richard Dawkins. This year Professor Sue Hartley delivers five presentations on ecology. rigb.org

>> JANUARY

▽ North American International Auto Show
Jan. 11–24, Detroit
Last year more than 650,000 people visited the NAIAS. This year 60 new vehicles will be premiered, making it one of the most eagerly anticipated technology events of the entire year. naias.com

>> FEBRUARY

>> Retech 2010
Feb. 3–5, Washington, D.C.
People interested in all renewable energy sectors — wind, solar, hydro, tidal, geothermal, biomass, biofuels, and waste-to-energy — gather in Washington to attend this event. Exhibitors will include equipment manufacturers and suppliers, systems providers, funders, and more. retech2010.com

>> Columbus Magi-Fest
Feb. 4–6, Columbus, Ohio
As MAKE, Volume 13 proved, makers love to make magic. Professional conjurers, magical hobbyists, prop makers, and other students of illusionism gather in Columbus to explore all aspects of their craft. magifest.org

>> Orange Blossom Special Star Party
Feb. 10–14, Dade City, Fla.
It's been exactly 400 years since Galileo first saw the moons of Jupiter. You can see them for yourself at the 16th installment of this astronomical gathering on the dark banks of the Withlacoochee River.
stpeteastronomyclub.org

>> Space Shuttle Atlantis STS-131 Launch
Feb. 11, Cape Canaveral, Fla.
It's the final year for the space shuttle program, so opportunities for live viewing of the 6 million pounds of fiery, smoky thrust are growing short. This launch will carry lab equipment to the International Space Station. kennedyspacecenter.com/events-launches.aspx

>> National High Magnetic Field Laboratory Open House
Feb. 27, Tallahassee, Fla.
Los Alamos National Laboratory, Florida State University, and the University of Florida open the doors of their world-class research laboratory for a public tour. Tour the record-breaking 45-tesla hybrid magnet and get a chance to meet and chat with Magnet Lab scientists. makezine.com/go/maglab

IMPORTANT: All times, dates, locations, and events are subject to change. Verify all information before making plans to attend.

Know an event that should be included? Send it to events@makezine.com. Sorry, it's not possible to list all submitted events in the magazine, but they will be listed online.

If you attend one of these events, please tell us about it at forums.makezine.com.

■ **We call it the Adventure Tower, and it's the** centerpiece of outdoor play for my two kids (ages 6 and 10) and all their friends. I designed and built the basic structure five years ago and have been adding (and sometimes subtracting) new play elements ever since. I treat it like my own giant Tinkertoy set, and have just as much fun building it as my kids have playing on it.

In the beginning, my goal was to build a big play structure that was rock solid (strong enough for adults to play on), had a small footprint, and maintained high play value as my kids grew.

Triangles are the answer to rock-solid structures, so the frame of the Adventure Tower is built around a tripod of 4×6 treated timbers 16 feet long. The base rail is made with three 12-foot timbers, and the deck uses three 8-foot timbers. The metal brackets are the only non-off-the-shelf items: an engineer friend designed the bracket that secures the peak of the tripod and the brackets that attach the horizontal 4×6 timbers to the tripod.

All the lumber and hardware were purchased at my local big-box home improvement store. The fun add-ons — the slide, rings, swing, driver's wheel, telescope, etc. — were purchased there as well. The cargo net came from an army surplus store, and the rock climbing handholds were purchased on the web.

I eventually built a clubhouse under the Adventure Tower, and the kids love it. The main door was "secretly" hidden on the climbing wall and I built two smaller secret escape doors. The kids pleaded for locks to be added to the doors, and I obliged and was told it made the clubhouse experience "way better!" Who would have thought?

I asked my 10-year-old son (a big MAKE magazine fan) what was most important to share in this article, and he said, "Tell them how big and strong it is." Size and toughness are important to him because he's played on the small, wimpy play structures in his friends' backyards.

In school last year, my son brought my scale model of the Adventure Tower in for show and tell, along with two little action figures. When I picked him up after school, he jumped in the car and said, "Everyone wants to come over to our house to play!" Now that's cool.

■ See photos and video of the Adventure Tower in use at adventuretower.com.

Tom Heck is a daddy, banjoist, team-builder, and maker.

Photograph by Tom Heck